T0205448

Predictive Computing and Information Security

P.K. Gupta · Vipin Tyagi
S.K. Singh

Predictive Computing and Information Security

P.K. Gupta
Department of Computer Science and Engineering
Jaypee University of Information Technology
Solan, HP
India

S.K. Singh
Department of Computer Science and Engineering
Indian Institute of Technology BHU
Varanasi, UP
India

Vipin Tyagi
Department of Computer Science and Engineering
Jaypee University of Engineering and Technology
Guna, MP
India

ISBN 978-981-13-5321-5 ISBN 978-981-10-5107-4 (eBook)
DOI 10.1007/978-981-10-5107-4

Softcover re-print of the Hardcover 1st edition 2017

Printed on acid-free paper

This Springer imprint is published by Springer Nature
The registered company is Springer Nature Singapore Pte Ltd.
The registered company address is: 152 Beach Road, #21-01/04 Gateway East, Singapore 189721, Singapore

Preface

Predictive Computing term is tossed with the advancements in computing field and with the evolution of computing techniques like Cloud computing, Pervasive computing, Internet of Things, Big data, etc. Predictions from these computing techniques are based on explanatory models and a good predictive model can be turned into a machine further. Predictive modeling encompasses much more than the uncovering patterns within data. The aim of predictive computing is to develop a model in a way that can predict accurately.

An important challenge related to data is its security. It becomes more crucial when the data is stored in the cloud as it raises the issues of security, privacy and trust. These issues become more crucial when we integrate cloud computing with Internet of Things, wireless sensor networks and other computing technologies.

The major objective of this book 'Predictive Computing and Information Security' is to bring forward the advancement and implementation of state-of-art techniques and approaches, design, development and innovative use of technologies for enhancing predictive computing while taking care of information security. Various recent algorithms, implementation techniques, and prediction techniques are discussed in the book.

Chapter 1 discusses the detailed view of predictive computing and presents the definition of predictive computing. This chapter also presents the various pillars of predictive computing that helps in digging of stored information for making sector specific predictions. Chapter 2 presents the detailed technical survey related to predictive computing and information security. It focuses on various hot key areas, where predictive computing based techniques and frameworks are applied to enhance human living, and environment. Key areas included for predictive framework are healthcare, smart home, navigation, e-commerce, etc. where technologies like cloud computing and Internet of Things could be implemented.

Chapter 3 discusses the detailed technical insight of predictive computing and represents the design architecture of predictive computing that includes various steps for designing a predictive model. This chapter also discusses about various predictive computing algorithms which are used to make accurate and effective predictions. Chapter 4 represents the integration of Internet of Things and cloud computing paradigm and discuses about the Internet of Things based Cloud centric architecture which is used for predictive analysis of physical activities of the users in sustainable health centers. Chapter 5 represents the various major challenges associated with Internet of Things based predictive computing techniques. Chapter 6 discusses various issues related to cloud security, privacy and trust, and presents a detailed overview of cloud threats and attacks. It also discusses a framework that would provide integrity of data to multiple users through Third Party Auditor (TPA) and also an algorithm to implement this framework over different types of clouds and implements the cryptographic algorithms, namely; RSA, Bcrypt and AES. Finally, Chapter 7 represents the various applications of predictive computing related to smart mobility, e-health, and e-Logistics domains. Real life applications of Predictive Computing are discussed in the book that will help readers in understanding the basic concept of the predictive computing.

We believe this book will be of interest to graduate students, educators and active researchers in academia, and engineers in industry who need to understand or implement predictive computing and information security. We hope this book will provide reference to many of the techniques used in the field as well as generate new research ideas to further advance the field.

This work would not have been possible without the help and mentoring from many individuals, including: Prof. Vinod Kumar, Vice Chancellor, JUIT, Wakngahat (HP), Prof. Samir Dev Gupta, Director & Academic Head, JUIT, Waknaghat (HP), Prof. J.S.P. Rai, Vice Chancellor, JUET, Guna (MP), Prof. V.K. Agarwal, Meerut (UP). We would also like to thank Prof. Mayank Singh, Prof. Vishwajeet Pattanaik, Prof. Shweta Suran, Prof. Sandeep Saxena, Prof. Umang Kant, and Prof. Krista Chaudhary for their continuous support in writing and organizing the things in required manner. The authors are sincerely thankful to the potential reviewers for their fruitful comments and suggestions to improve the quality of this book. Authors are also thankful to publication partner and authorities, Ms. Suvira Srivastav, Ms. Yeshmeena Bisht, Ms. Nidhi Chandhoke, and Ms. Krati Shrivastav for their kind support.

Solan, India P.K. Gupta
Guna, India Vipin Tyagi
Varanasi, India S.K. Singh

Contents

Acronyms

AES	Advanced Encryption Standard
ANP	Analytic Network Process
ANS	Autonomic Navigation System
API	Application Programming Interface
BAN	Body Area Network
BI	Business Intelligence
CADS	Continuous Authentication on Data Streams
CCAF	Cloud Computing Adoption Framework
CI	Computational Intelligence
CIA	Confidentiality, Integrity, and Availability
CMED	Cloud based MEDical system
CPU	Central Processing Unit
CS	Cloud Server
CSA	Cloud Service Architecture
CSP	Cloud Service Providers
DCTP	Data Collection based on Trajectory Prediction
DDOS	Distributed Denial of Service
DEMATEL	Decision-Making Trial and Evaluation Laboratory
DES	Data Encryption Standard
DNA	Deoxyribo Nucleic Acid
DoS	Denial of Service
DSA	Digital Signature Algorithm
DSMS	Data Stream Management System
ECG	Electrocardiogram
e-CLV	Electronic Customer Lifetime Value
E-commerce	Electronic Commerce
EDOS	Economic Denial of Service
e-Governance	Electronic Governance
EHR	Electronic health Records
ELS	E-commerce Logistics System

E-RS	Enhanced Reed-Solomon
GB	Gigabyte
GDP	Gross Domestic Product
GPS	Global Positioning System
GSM	Global System for Mobile communication
HMM	Hidden Markov Model
IA	Information Accountability
IaaS	Infrastructure as a Service
ICT	Information and Communication Technology
IoT	Internet of Things
IoV	Internet of Vehicle
ISG	Information Security Governance
IT	Information Technology
ITSC	Intelligent Traffic Control System
MCDM	Multi criteria Decision Making
MD	Message Digest
MDP	Multi-Dimensional Password
ML	Machine Learning
MLA	Multi-level Authentication Scheme
MLC	Multi-level Cryptography
MOLA	Multi Objective Learning Automata
P2P	Peer to Peer
PaaS	Platform as a Service
PARAMO	PARAllel predictive MOdelling
PB	PetaByte
PCC	Predictive Cloud Computing
PccP	Preserving Cloud Computing Privacy
PHMS	Patient Health Monitoring System
PIoT	Predictive Internet of Things
PMC	Predictive Mobile Computing
POR	Proof of Retrievability
PPCA	Probabilistic Principal Component Analysis
PRDC	Permission-based RFID Data Collection Algorithm
PTT	Prediction of Trigger time
PoD	Prediction of Distance
QoS	Quality of Service
RBAC	Role Based Access Control
RBE	Role Based Encryption
RC	Rivest Cipher
RDIC	Remote Data Integrity Checking
REF	Reference
RFID	Radio Frequency Identification
RPM	Remote Patient Monitoring
RSA	Rivest Shamir Adleman
SaaS	Software as a Service

SAMOA	Socially Aware and Mobile Architecture
SBA	Seed Block Algorithm
SDG	Stochastic Gradient Descent
SecCTP	Secure Cloud Transmission Protocol
SHA	Secure Hash Algorithm
s-Health	Smart Health
SLA	Service Level Agreement
SOAP	Simple Object Access Protocol
SQL	Structured Query Language
SSL	Secure Socket Layer
SWF	Smallest Window First
TB	Terabyte
TLS	Transport Layer Security
TPA	Third Party Auditor
TTP	Trusted Third Party
UML	Unified Modeling Language
V2V	Vehicle-to-Vehicle
VANET	Vehicular ad-hoc network
VM	Virtual Machines
VPN	Virtual Private Network
Wi-Fi	Wireless Fidelity
WSDL	Web Services Description Language
WSN	Wireless sensor networks
XML	eXtensible Markup Language

List of Figures

List of Tables

About the Authors

Dr. P.K. Gupta, a Post-Doctorate from the University of Pretoria, South Africa (2015–2016) in the Department of Electrical, Electronic and Computer Engineering, is currently an Assistant Professor (Sr. Grade) at Jaypee University of Information Technology (JUIT), Himachal Pradesh (HP), India. He has more than 15 years of national and international experience in the Information Technology (IT) industry and academics. He has authored a number of research papers in peer-reviewed international journals and conferences. Further, Dr. Gupta is an Associate Editor of IEEE Access. His research areas include Internet of Things, Cloud Computing, Sustainable Computing and Storage Networks.

Prof. Vipin Tyagi Fellow-IETE, is currently working as a Professor in Computer Science and Engineering Department and Head—Faculty of Mathematical Sciences at Jaypee University of Engineering and Technology (JUET), Madhya Pradesh (MP), India. He is Vice President of the Computer Society of India, Region 3, and is associated with the Society's Special Interest Group on Cyber Forensics. He was President of the Engineering Sciences Section of the Indian Science Congress Association for the term 2010–2011. He has published a number of papers in various reputed journals and advanced research series, and has attended several national and international conferences. He is an expert in the area of Cyber Security, Cyber Forensics and Image Processing.

Prof. Sanjay Kumar Singh is a Professor in the Department of Computer Science and Engineering, Indian Institute of Technology (IIT), Uttar Pradesh (UP), India. He has been certified as a Novell Engineer (CNE) and Novell Administrator (CNA) by Novell Netware, USA. He is a member of LIMSTE, the IEEE, International Association of Engineers and the ISCE. He has over 70 national and international journal publications, book chapters and conference papers to his credit. His research areas include Biometrics, Computer Vision, Image Processing, Video Processing, Pattern Recognition and Artificial Intelligence.

Chapter 1
Introduction to Predictive Computing

1.1 Introduction

The existing computing techniques are becoming more challenging to compete with the global development. However, the modern computing techniques have more focus on interdisciplinary approaches to perform complex tasks and development to satisfy human needs. The computing techniques designed to strengthen communication among the living and nonliving objects, to reduce energy consumption and time by applying smart methods for predicting navigation paths, prediction of health diseases and monitoring, weather prediction, etc., can be classified under predictive computing techniques. Further, the computing techniques like Green Computing, Ubiquitous Computing, Internet of Things (IoT), Human–Computer Interaction, and Intelligent Transportation are considered as a few major techniques which perform computations in real time or near real time. Table 1.1 presents the evolution of computing paradigm and from which we observe the shift of programming paradigm that was completely based on desktop computing since 1960–1990. During this period, the procedural languages [1–8] like ALGOL [4, 5], FORTRAN [6, 7], Pascal [7, 8] and C [7] were the common programming languages for designing and developing computer programs. As the advancements in technology have taken place, the evolution of other programming languages like functional programming [2, 3], and declarative programming languages [2, 3] like LOGO, PROLOG [9, 10] and ASP [11] has taken over the procedural programming languages. With the significant technological change after 1990 and emerging of object oriented programming languages [2] like Java, C#, etc., the computing scenario has changed completely and new methods are being introduced. With the use of Internet technology, significant developmental changes have taken place with the help of web-based and scripting languages [12] in the market for designing of websites and Internet-based systems. Web-based computing has changed the way of use of computer systems and this has led to the development of applications that can work parallel or concurrent [13–17], agent-based

P.K. Gupta et al., *Predictive Computing and Information Security*,
DOI 10.1007/978-981-10-5107-4_1

Table 1.1 Evolution of computing paradigm

S. no.	Computing paradigm	Programming language	Era of use
1.	Procedural/Imperative programming [1–3]	ALGOL [4, 5]	1958–1968
		FORTRAN and its variants [6, 7]	1950–till now
		Pascal and its variants [7, 8]	1970–2012
		C and its variants [7]	1972–2011
2.	Functional [2, 3]	LISP and its variants	1958–2013
		ML or MIRANDA	1985–1989
3.	Declarative [2, 3]	LOGO or PROLOG and its variants [9, 10]	1972–1995
		Answer set programming (ASP) [11]	1993–1999
4.	Object oriented [2]	C++	1983–till now
		JAVA and its variants	1995–till now
		C# and its variants	2000–till now
		SMALLTALK and its variants	1972–1990
		EIFFEL and its variants	1986–2009
5.	Scripting [12]	VBScript and its variants	1996–till now
		JavaScript and its variants	1995–till now
6.	Parallel/Concurrent [13, 14]	Ada [15]	1982–till now
		Erlang [16]	1986–till now
		RUST [17]	2010–till now
7.	Agent based [18, 19]	AgentSpeak [20]	1996–till now

environment [18–20]. These significant changes in computing and communication have eased down the connectivity of the devices and data processing could become possible in real time to predict new findings like prediction of change detection in ground images, prediction of shortest navigation path of a vehicle by using existing driving habits and prediction of health status, etc. These changes and advancements in computing field have opened the doors to information security related issues and problems. The use of sensor technologies, Internet technologies, virtualization and real-time computation all along the way has introduced new bugs and vulnerabilities in the designed system. However, a variety of information security techniques are proposed over the period of time which can be used to make communication secure.

In [21], Bartels has proposed a new kind of computing where convergence of innovations in software architecture, backend operations, communications and to various client devices connected to the network lets advance computing technology work together to find and solve the complex business problems in innovative manner that could not be addressed by the last generation computing techniques. Seed of '*innovations*' adds new capabilities to existing technologies for real-time situational awareness and automated analysis. Researchers and software developers are introducing this seed to their software, hardware and communications to solve the complex business problem smartly.

1.2 Definitions

The term 'Predictive Computing' is tossed with the advancements in computing field and with the evolution of computing techniques like cloud computing, pervasive computing, IoT, big data, etc. Predictions from these computing techniques are based on explanatory models and a good predictive model can be turned into a machine. In this section, we have presented the general definition of computing, and few definitions related to the concept of predictive computing and Predictive Analytics.

- Computing
  Over the period of time, a wide range of definitions has been given related to terms '*Computing*' and '*Predictive Analytics*'. Comer et al. [22] have stated that '*computing is the systematic study of algorithmic processes that describe and transform information: Their theory, analysis, design, efficiency, implementation, and application*'. In [23], Denning has stated that 'computing is a natural science' as the term '*computation*' and '*information processes*' are existing in the literature long ago even before the invention of computers. In [24], Shackelford et al. stated, '*computing is any goal-oriented activity requiring, benefiting from, or creating computers and includes designing and building of hardware and software systems for a wide range of purposes; processing, structuring, and managing various kinds of information; doing scientific studies using computers; making computer systems behave intelligently; creating and using communications and entertainment media; finding and gathering information relevant to any particular purpose, and so on*'.
- Predictive Analytics
  Today's businesses use variety of business models which in turn are driven by data analytics which provides useful methodology for exploring available data and for developing significant models to serve the purpose of an entity [25]. A variety of terms can be found related to analytics like business analytics, academic analytics, learning analytics, predictive analytics, etc. In [26], Matt has stated that analytics means different things to different peoples and is not a one-size-fits-all endeavour. In [27], Eckerson stated that '*Predictive analytics is*

a set of business intelligence (BI) technologies that uncovers relationships and patterns within large volumes of data that can be used to predict behavior and events. Unlike other BI technologies, predictive analytics is forward-looking, using past events to anticipate the future'. In [28], IBM stated that '*Predictive analytics connects data to effective action by drawing reliable conclusions about current conditions and future events*'. According to SPSS '*Predictive analytics is both a business process and a set of related technologies that leverages an organization's business knowledge by applying sophisticated analysis techniques to enterprise data*' [29].

- Predictive Computing
  According to Nadin [30], '*Predictive computation is the path that leads from reaction-based forms of computing to anticipatory forms of computing*'. 'To predict' means to state something about next step or next sequence. The predictive value could be find with respect to time, space, words, expression in language, degree or significance whereas, '*Predictions can be time-independent (extenders), pertinent to simultaneous occurrences (portents), or can infer from data describing a previous state or the current state of the world to a future state*'. In [31, 32], Frost & Sullivan have stated that IoT market will continue its growth in future and architecture of IoT 2.0 will enable self-healing events in the connected system. IoT 2.0 is supposed to react to various events, to using *sentient tools and cognition or 'predictive computing*'. Thus, a definition of predictive computing can be proposed here:

Predictive computing presents an algorithmic approach that processes collected data of living and nonliving entities periodically from different sensor nodes in the network to develop an effective prediction model where the next step, or next sequence, or the future state of the system/user activity can be represented effectively.

Here, we have used the term 'model' in general sense which refers to abstraction of system that can derive some knowledge after processing the collected data. We can realize the existence of predictive computing from the following scenario (Fig. 1.1) that represents the growing trends of business transactions over use of computing.

1.3 Pillars of Predictive Computing

In [33], Stankovic has represented that by using the techniques like pervasive computing, Internet of Things, wireless sensor networks (WSN), mobile computing and cyber-physical systems, the world can be transformed into smart world. These computing techniques provide the basis for predictive computing. The author has

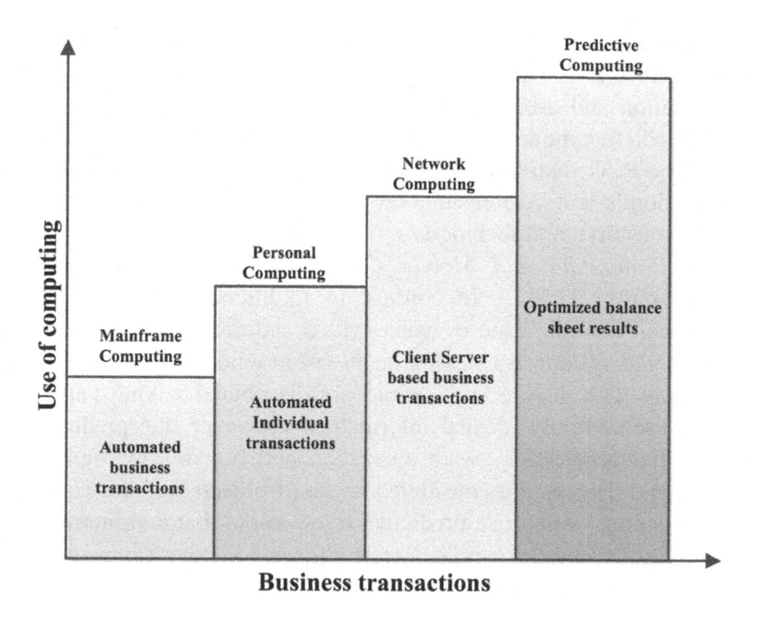

Fig. 1.1 Role of predictive computing in business transaction

also focused on the major research problems like security, privacy, massive scaling, architecture, and robustness, etc., for creating smart applications that can make predictions to make our lives easier and comfortable. Figure 1.2 represents the six core computing techniques that can act as six pillars for predictive computing.

- *Predictive Computing and Internet of Things*—also known as predictive internet of things (PIoT). The architecture of IoT 2.0 will enable self-healing events and is supposed to react to various events related to living and nonliving objects in the network by making use of predictive computing [31, 32]. Connected IoT nodes in the network are used to collect the information related to the object which is further stored in the database or clouds.

Fig. 1.2 Six pillars of predictive computing

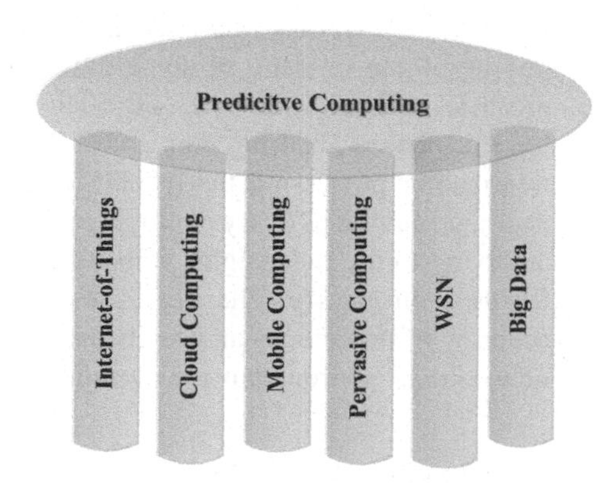

- *Predictive Computing and Cloud Computing*—also known as predictive cloud computing (PCC). The major objective of PCC is to develop and provide a smart allocation and deallocation of servers by combining ensembles of forecasts and predictive modeling to determine the future origin demand for website content [34]. PCC distributes the load evenly over various virtual machines located in clouds and also ensures the security of data by using various PCC security frameworks and techniques.
- *Predictive Computing and Mobile Computing*—also referred as predictive mobile computing (PMC). In contrast to traditional telephony, the mobile computing can communicate by voice, video and data wirelessly over diversified devices and systems by making the use of mobile communication standards and protocols. The mobile applications are distributed among various mobile nodes and use centrally located information. However, the predictive mobile computing has application-aware adaptation and has an environment-sensing ability. Mobile devices are considered as smart objects and are equipped with smart applications which use predictive frameworks that continuously monitor user's activity on the devices and send information to various servers for its further processing and making predictions [35, 36].
- *Predictive Computing and Pervasive Computing*—also known as smart computing and considered as a successor of mobile computing and ubiquitous computing. As pervasive computing embeds the computational capability into day-to-day objects to make them effectively communicate, predictive pervasive computing makes devices and objects smart and intelligent that can take or provide decisions on the basis of available information without any user intervention. Devices involved for smart computing are computers, laptops, tablets, smart phones, sensors, wearable devices, etc. [36].
- *Predictive Computing and Wireless Sensor Network* (*WSN*)—wireless sensor network is the backbone for implementing the predictive computing techniques and frameworks. The advancements in computing world have only become possible because of WSN, which Wireless sensor networks consist of geographically distributed autonomous devices using sensor to monitor physical and environmental conditions. The sensors used in WSN are one of the smart objects that continuously monitor and forward the information between user and data centres.
- *Predictive Computing and Big Data*—The sensor nodes used in WSN generate heavy volume of data. In some cases, daily collection of data reaches up to gigabyte (GB) and terabyte (TB). This collection of data supports creation of huge database called big data and demands cost effective, innovative forms of information processing that enable enhanced insight, process automation and decision-making [37]. By deploying predictive models over this big data, people are trying to use the insights gained from big data to uncover new opportunities for their businesses. The big data plays an important role in improving the accuracy of predictions and can be used in health care, tourist, agriculture, social networking, environment monitoring, etc. [38].

1.4 Horizons of Predictive Computing

The existence of predictive computing could be in every major sector as shown in Table 1.2. Here, it can be observed that predictive computing plays a vital role everywhere, from smart home to healthcare sector. The integration of IoT, cloud computing and wireless sensor networks has made it possible to find the predicted result in real time for different areas. The predictive computing consists of various smart objects connected via wireless sensor network for collection of the data which gets stored in clouds for further its processing. Possibilities of predictive computing span over health care, transport, travel, sales, smart home like other many sectors (Table 1.2).

According to Gartner's hype cycle of 2015 [13], for emerging technologies IoT, mobility, digital business and analytics will play a lead role in growth of opportunities and provide the new experiences to the customers and organizations. This megatrend of hype cycle is shown in Fig. 1.3.

There will be a significant growth in the area of digital marketing and digital business. The current marketing trends will be replaced and reinvented with predictive applications solutions. These applications will be able to analyse the human behaviour from the existing data and will provide the solutions intelligently and effectively.

1.5 Role of Information Security and Techniques

The role of information security is becoming vital at each level of communication, storage and user access. Integration of technologies like IoT, cloud computing and wireless sensor networks involves billions of nodes and devices generate data that need to be stored with virtual machines located in cloud, communication between these devices and user requires access control mechanism and communication mechanisms are intact.

Figure 1.4 represents the major security-related challenges associated to these areas and application security framework. On the basis of object identification in predictive computing technology, system level characteristics could be categorized where predictive parameters can be identified, and communicated with each other. Based on this consideration, the research areas like computing and communication techniques, interconnected systems and distributed intelligence have been identified. Devices enabled with smart computing must be able to identify, perform computing and communication with each other. RFID (Radio Frequency Identification) technique is becoming more popular for identification of objects, whereas low-power research communication techniques are being used in sensor networking [39, 40]. In future, low-power and low-cost communication technologies will be considered for communication purpose among smart objects for prediction purpose. Interconnected systems play the important role in implementation

Table 1.2 Horizons of predictive computing

	Areas	Uses					
Predictive computing	Smart home	Entertainment	Security	Kitchen	Home appliances		
	Transport	Navigation	Parking	Logistics	Traffics	Emergency services	
	Community	Factory	Retail	Surveillance	Business intelligence	Smart metering	Environment
	National	Infrastructure	Smart grid	Weather forecasting	Defence	Remote monitoring	Population
	Industries	Location	Security	Surveillance	Audit		
	Policy makers	Health	Water	Electricity	Population	Education	Goods and services
	Personal user	Handheld devices	Desktops	Banking			
	Health care	Patients monitoring	Emergency	Records	OPD		

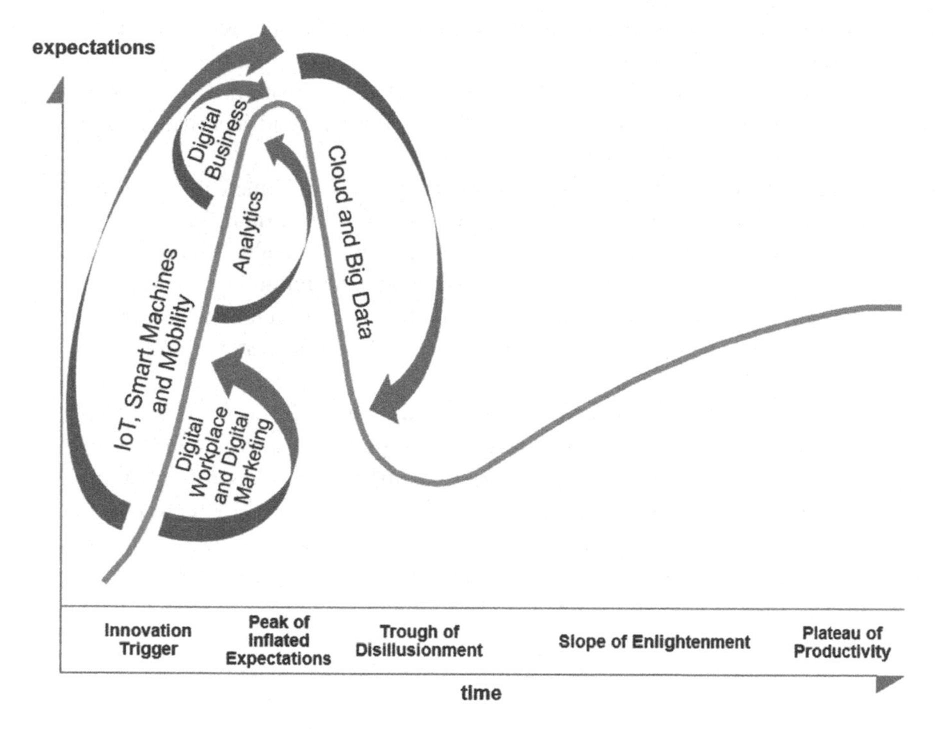

Fig. 1.3 Megatrends of significant technologies [13]

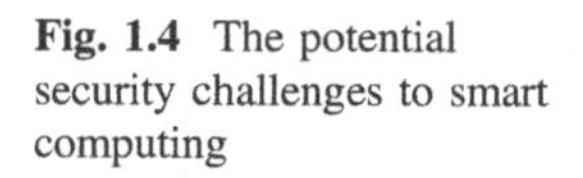
Fig. 1.4 The potential security challenges to smart computing

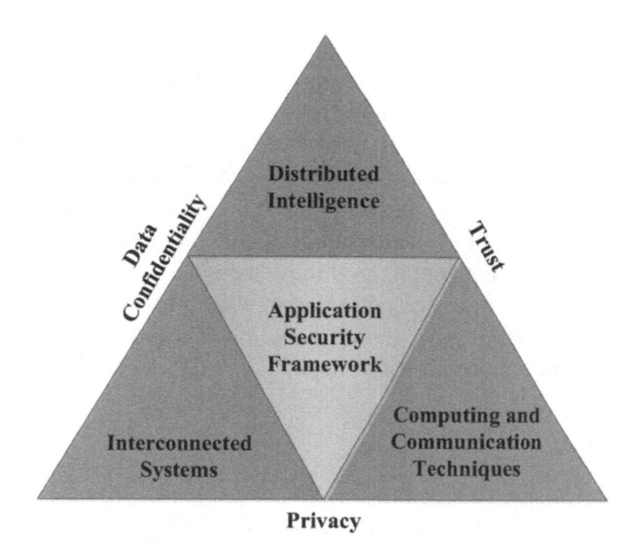

of smart objects. Interconnection of these smart objects involves many challenges from network design to communication. The research directions in distributed intelligence get established once the system based on smart objects is interconnected. One of the challenging tasks is to handle, interpret and security of incoming data from these smart objects [41].

Security, in context to predictive computing, can be defined in terms of data confidentiality, privacy, and trust. For ensuring data confidentiality, we must ensure that the access control mechanism of data is assigned to right person or relevant bodies. Any breach in confidentiality will result into failure of overall system. Privacy defines the rules on data to be accessed by individual personnel. Since predictive computing is integrated with other technologies at various levels, so privacy of data must be maintained at each level of communication. Wireless channels and remote access technologies for data expose the overall system for risks and threats. Therefore, the privacy represents an open issue for system development and trust refers to designed or implemented security and privacy policies for handling of access control and to resource required by them [42]. Trust must be established between both parties, one who requires the resources and another who provides the credentials. A summary of various techniques proposed by various authors related to data confidentiality, privacy and trust is given in Table 1.3.

1.6 Summary

The advancement in technology has brought significant changes into existing products that replace the current framework with new kind of intelligent frameworks. These new frameworks are able to predict the information in real time or near real time with any intervention from user of the system. For making prediction, technologies like cloud computing, IoT, pervasive computing, mobile computing, etc., are playing important role in relation to wireless sensor network. In this chapter, we have discussed predictive computing and the insights of risks associated with the integration of these technologies. The role of information security and techniques has been presented to ensure the data confidentiality, privacy, and trust for communication of data, handling of user control mechanism at various levels to provide secure access to data.

Table 1.3 Techniques for secure trusted computing

S. no.	Security challenge	Author(s)	Proposed/Applied technique	Technical description	Limitations and challenges
1	Data confidentiality	Sandhu et al. [43]	Role-based access control (RBAC)	Focused on multi-user and multi-application online systems and created various roles for different types of work roles	– Development of systematic methodology – Constraints in context to RBAC – Management aspects of RBAC
2		Papadopoulos et al. [44]	Reference (REF), Continuous authentication on data streams (CADS)	Reduced processing cost and communication overhead between server and clients. Extended for multiple users	– Performance of system – Single verification process – Handling spatial queries
3		Ali et al. [45]	FT-RC4	Security architecture for data stream systems and focuses on data confidentiality and integrity	– Evaluation of proposed scheme – Integration with other technologies
4		Nehme et al. [46]	Data stream management system (DSMS)	Access control policies get embedded into data streams using security constraints	– Incremental access control policies – Runtime changes in role assignments and their effect on query processing
5		Hammad et al. [47]	Smallest window first (SWF), greedy algorithm	Focused on shared execution of multiple window join queries over data streams	– Additional memory overhead – Complex nature of SWF
6		Hu and Evans [48]	SAWN	A protocol for secure data aggregation to single node in wireless network	– Two consecutive nodes cannot be compromised – Delayed verification
7		Bagga et al. [49]	SEDAN	A protocol provides secure data aggregation for wireless sensor networks and verification mechanism for data integrity	– Tree communication topology – Considered SAWN assumption of two consecutive nodes
8	Privacy	Van Lamsweerde and Letire [50]	Temporal obstacle analysis	Formal technique for reasoning about obstacles to the satisfaction of goals, requirements, and assumptions	– Implementation of technique – Cost of its resolution
9		Liu et al. [51]	*i** modeling framework	Focused on security issues about relationships among social actors	– Security examined for P2P domain – Functional vulnerabilities

(continued)

Table 1.3 (continued)

S. no.	Security challenge	Author(s)	Proposed/Applied technique	Technical description	Limitations and challenges
10		Mouratidis et al. [52]	Tropos	Integration of security and system engineering in system development	– Application on different processes – Formal evaluation
11		Kalloniatis et al. [53]	PriS	A security method included during early phase of system development	– Use of EKD method – Need for automated tool support
12		Coen-Porisini et al. [54]	UML model for privacy policy	UML conceptual model to represent the privacy policy for privacy aware systems	– Concrete implementations – Scalability
13	Trust	Ren et al. [55]	Modified distributed trust establishment approach	Established trust initialization in ad hoc networks	– Testing with mobile ad hoc networks – Performance issue
14		Liang and Shi [56]	PET, M-CUBE	Established trust for high-level resource management in P2P networking	– Uses PKI for public key distribution – Implements greedy approach – Unique and stable ID for each pair
15		Blaze et al. [42]	Policy maker	General framework can be applied to any service wherever cryptography is required	– Restrictions on predicate uses – Implementation of prototype in different applications

References

1. Ishida T, Sasaki Y, Fukuhara Y (1991) Use of procedural programming languages for controlling production systems. Paper presented at seventh conference on artificial intelligence applications, Miami Beach, FL, USA
2. Hagen W (2006) The style of sources: remarks on the theory and history of programming languages. In: Wendy H, Thomas K (eds) New media, old media. A history and theory reader. Taylor and Francis group, Routledge, pp 157–174
3. Rajaraman V (1998) Programming languages. Resonance 43–54
4. Naur P et al (1997) Revised report on the algorithmic language Algol 60. In: Proceedings of ALGOL-like languages, Birkhäuser, Boston, pp 19–49
5. van Wijngaarden A (1981) Revised report of the algorithmic language Algol 68. Proc ALGOL Bull (Sup 47):1–119
6. Backus J (1978) The history of Fortran I, II, and III. In: Proceedings of history of programming languages I, ACM, pp 25–74
7. Holtz NM, Rasdorf WJ (1988) An evaluation of programming languages and language features for engineering software development. Eng Comput 3(4):183–199
8. Wirth N (1971) The programming language Pascal. Acta Informatica 1(1):35–63
9. Abelson H, Goodman N, Rudolph L (1971) Logo manual. Artificial Intelligence Lab, Available via DSpace@MIT. https://dspace.mit.edu/handle/1721.1/6226. Accessed 25 May 2017
10. Brusilovsky P, Calabrese E et al (1997) Mini-languages: a way to learn programming principles. Educ Inf Technol 2(1):65–83
11. Bonatti P, Calimeri F, Leone N, Ricca F (2010) Answer set programming. A 25-year perspective on logic programming, pp 159–182
12. Ousterhout JK (1998) Scripting: higher level programming for the 21st century. Computer 31(3):23–30
13. Burton B, Willis DA (2015) Gartner's hype cycles for 2015: five megatrends shift the computing landscape. Recuperado de. https://www.gartner.com/doc/3111522/gartners--hype--cycles--megatrends--shift. Accessed 12 Feb 2017
14. Løvengreen HH (2004) Processes and threads. Course notes. Available via IMM. http://www2.compute.dtu.dk/courses/02158/proc.pdf. Accessed 15 Mar 2017
15. Burns A, Wellings A (1995) Concurrency in ADA. Cambridge University Press, Cambridge
16. Armstrong J (2010) Erlang. Commun ACM 53(9):68–75
17. Matsakis ND, Klock FS II (2014) The rust language. ACM SIGAda Ada Lett 34(3):103–104
18. Hu C, Mao X, Li M, Zhu Z (2014) Organization-based agent-oriented programming: model, mechanisms, and language. Front Comput Sci 8(1):33–51
19. Dennis LA, Fisher M, Webster MP, Bordini RH (2012) Model checking agent programming languages. Autom Softw Eng 19(1):5–63
20. Rao A (1996) AgentSpeak(L): BDI agents speak out in a logical computable language. In: Proceedings of 7th European workshop on modelling autonomous agents in a multi-agent world (MAAMAW). 1038, pp 42–55
21. Bartels AH (2009) Smart computing drives the new era of IT growth. Forrester Inc., Cambridge, pp 1–44
22. Comer DE, Gries D, Mulder MC, Tucker A, Turner AJ, Young PR, Denning PJ (1989) Computing as a discipline. Commun ACM 32(1):9–23
23. Denning PJ (2007) Computing is a natural science. Commun ACM 50(7):13–18
24. Shackelford R, McGettrick A, Sloan R, Topi H, Davies G, Kamali R, Cross J, Impagliazzo J, LeBlanc R, Lunt B (2005) Computing curricula 2005: the overview report. ACM SIGCSE Bull 38(1):456–457
25. Hawkins BL (2008) Accountability, demands for information, and the role of the campus IT organization. The tower and the cloud, pp 98–104

26. Matt D (2015) Analytics: meaning different things to different people. http://www.itproportal.com/2015/10/23/analytics-meaning-different-things-to-different-people/. Accessed 15 May 2017
27. Eckerson WW (2007) Predictive analytics. Extending the value of your data warehousing investment. TDWI Best Practices Report, Q1, pp 1–36
28. IBM (2010) Real world predictive analytics. Available via IBM Software. http://www.alemsistem.ba/media/1592/wp_real-world-predictive-analytics_ibm_spss.pdf. Accessed 10 Feb 2017
29. SPSS (2010) Predictive analytics: defined. http://www.spss.com.hk/corpinfo/predictive.htm. Accessed 10 Feb 2017
30. Nadin M (2016) Predictive and anticipatory computing. In: Phillip AL (ed) Encyclopaedia of computer science and technology, 2nd edn. CRC Press, pp 643–659
31. Frost & Sullivan (2017) Next evolution in IoT will be sentient tools and cognition or predictive computing. http://www.prnewswire.com/news-releases/next-evolution-in-iot-will-be-sentient-tools-and-cognition-or-predictive-computing-300450448.html. Accessed 10 Feb 2017
32. Frost & Sullivan (2016) IoT 2.0: predictive computing. https://ww2.frost.com/frost-perspectives/iot-20-predictive-computing. Accessed 10 Feb 2017
33. Stankovic JA (2014) Research directions for the internet of things. IEEE Internet Things J 1 (1):3–9
34. Baughman AK, Bogdany RJ, McAvoy C, Locke R, O'Connell B, Upton C (2015) Predictive cloud computing with big data: professional golf and tennis forecasting. IEEE Comput Intell Mag 10(3):62–76
35. Liu GY, Maguire GQ (1995) A predictive mobility management algorithm for wireless mobile computing and communications. In: Proceedings of fourth international conference on universal personal communications, Tokyo, Japan, IEEE, pp 268–272
36. Rajkamal (2007) Mobile computing. Oxford University Press, Oxford
37. Thota C, Manogaran G, Lopez D, Vijayakumar V (2016) Big data security framework for distributed cloud data centers. Cybersecurity breaches and issues surrounding online threat protection, IGI global, pp 288–310
38. Song H, Liu H (2017) Predicting tourist demand using big data. In: Analytics in smart tourism design. Springer International Publishing, Berlin
39. Enz CC, El-Hoiydi A, Decotignie JD, Peiris V (2004) WiseNET: an ultralow-power wireless sensor network solution. Computer 37(8):62–70
40. Lu G, Krishnamachari B, Raghavendra CS (2004) An adaptive energy-efficient and low-latency MAC for data gathering in wireless sensor networks. In: Proceedings of 18th international parallel and distributed processing symposium, Mexico, pp 62–70
41. Kranz M, Holleis P, Schmidt A (2010) Embedded interaction: interacting with the internet of things. IEEE Internet Comput 14(2):46–53
42. Blaze M, Feigenbaum J, Lacy J (1996) Decentralized trust management. In: Proceedings of symposium on security and privacy, Oakland, CA, USA, pp 164–173
43. Sandhu RS, Coyne EJ, Feinstein HL, Youman CE (1996) Role-based access control models. Computer 29(2):38–47
44. Papadopoulos S, Yang Y, Papadias D (2007) CADS: continuous authentication on data streams. In: Proceedings of 33rd international conference on very large data bases, Vienna, Austria, pp 135–146
45. Ali M, ElTabakh M, Nita-Rotaru C (2005) FT-RC4: a robust security mechanism for data stream systems. Comput Sci Tech Rep Paper 1638:1–10
46. Nehme RV, Rundensteiner EA, Bertino E (2008) A security punctuation framework for enforcing access control on streaming data. In: Proceedings of 24th international conference on data engineering (ICDE 2008), Cancun, Mexico, pp 406–415
47. Hammad MA, Franklin MJ, Aref WG, Elmagarmid AK (2003) Scheduling for shared window joins over data streams. In: Proceedings of 29th international conference on very large data bases, vol 29. Berlin, Germany, pp 297–308

48. Hu L, Evans D (2003) Secure aggregation for wireless networks. In: Proceedings of symposium on applications and the internet workshops, Orlando, FL, USA, pp 384–391
49. Bagaa M, Lasla N, Ouadjaout A, Challal Y (2007) Sedan: secure and efficient protocol for data aggregation in wireless sensor networks. In: Proceedings of 32nd conference on local computer networks (LCN 2007), Dublin, Ireland, IEEE, pp 1053–1060
50. Van Lamsweerde A, Letier E (2000) Handling obstacles in goal-oriented requirements engineering. IEEE Trans Softw Eng 26(10):978–1005
51. Liu L, Yu E, Mylopoulos J (2002) Analyzing security requirements as relationships among strategic actors. In: Proceedings of symposium on requirements engineering for information security (SREIS'02), Raleigh, North Carolina, pp 1–14
52. Mouratidis H, Giorgini P, Manson G, (2003) Integrating security and systems engineering: towards the modelling of secure information systems. In: International conference on advanced information systems engineering, Klagenfurt, Austria, pp 63–78
53. Kalloniatis C et al (2008) Addressing privacy requirements in system design: the PriS method. Requir Eng 13(3):241–255
54. Coen-Porisini A, Colombo P, Sicari S (2010) Privacy aware systems: from models to patterns. IGI Global, pp 232–259
55. Ren K, Li T, Wan Z, Bao F, Deng RH, Kim K (2004) Highly reliable trust establishment scheme in ad hoc networks. Comput Netw 45(6):687–699
56. Liang Z, Shi W (2005) Enforcing cooperative resource sharing in untrusted P2P computing environments. Mob Netw Appl 10(6):971–983

Chapter 2
Predictive Computing and Information Security: A Technical Review

2.1 Introduction

Future of computing depends upon the effective integration of existing technologies and computing techniques. Integration of cloud computing and IoT provides a new dimension to computing known as 'Predictive Computing' and opens the new possibilities to researchers and developers. Predictive computing utilizes the data to make real-time or near real-time predictions for making life easier and comfortable. The predictive computing consists of various smart objects connected via wireless sensor network for collection of the data which gets stored in clouds for further its processing. The possibilities of predictive computing spans over healthcare, transport, travel, sales, smart home like other many sectors. With the rising demand of sensor-based products and applications for making predictions in many sectors also include the security risks and privacy issues related to data collection and data storage.

Use of predictive computing is getting popular as it is possible to design a smart environment, capable of monitoring air and water pollution, prediction of weather, earthquakes, detecting forest fires, tsunamis and various types of disasters so that early measures could be taken to reduce their devastating effects. Further, to make tourism sector attractive, the travel industries have come out from existing practice of making prediction based on travellers surveys and expert views to real-time collection of information of travellers location and updates about location in terms of photos, nearby spots and many more. Moreover, the big data analysts can capture the information from various photos posted on Facebook and other social networks [1–3]. With the increased use of IoT, a number of smart home appliances have been released into the market that include smart TV, smart refrigerator, smart lights, smart cooling and heating devices that can be controlled by mobile devices or desktops from remote location using internet [4]. Similarly, the use of sensor-based IoT devices is rapidly increasing in predicting the shortest path for vehicle navigation and avoiding of traffic congestion. Vehicle route is predicted in real-time by

P.K. Gupta et al., *Predictive Computing and Information Security*,
DOI 10.1007/978-981-10-5107-4_2

using the existing driving habits in the database and GPS to find the traffic congestion or road block like scenarios [5, 6]. A variety of devices is available in the market to monitor the health of a person. Monitoring of health, using smart blood pressure monitor, glucose monitor, pulse monitoring device, tracking of user activity running, walking, heartbeat, etc., can be done using various sensors and storing of this information over cloud for its further monitoring by healthcare personnel [7]. The healthcare personnel can make the prediction on the basis of received data and can provide suggestion or medications accordingly. In this chapter, we have discussed various such predictive frameworks and computing techniques based on IoT and cloud computing to perform various predictions. As the dependency on these smart devices is increasing and these smart objects are becoming the soft target of attackers to breach the user's privacy. Managing data confidentiality, integrity is becoming a major task it is because a number of restrictions associated with these smart devices in terms limited battery power, lack of complex encryption algorithm, poor access control at user level and lack of secure communication methods, etc. This chapter discusses various information security techniques and frameworks to maintain the confidentiality, integrity, availability and trustworthiness of data. We have listed various available security threats and their countermeasures with respect to two major technologies cloud computing and IoT.

2.2 Google Trend Analysis

With the growing shift towards the use of IoT, cloud computing and predictive computing, we have tried to find and analyze the Google search trends for computing terms like 'Cloud Computing', Internet-of-Things, 'Predictive analytics' and 'Information Security' from year 2008 to 2017 (up to May 2017).

The Google search trend results are shown in Fig. 2.1 and it can be observed that cloud computing and information security are two major areas which lead the global market whereas significant advancements are taking place with IoT and predictive analytics. The authors have also tried to find the trend results for 'Predictive computing' but no results are returned for the same. Market role of predictive computing is gearing up and one can expect a big demand and growth in future with the increased use of IoT 2.0 [8]. We have also analyzed the search trends at region level which are represented in Fig. 2.2 for all the previously mentioned computing techniques. From these results, it is clear that the regions like Singapore, India, Kenya and Uganda are the major regions, related to use of mentioned buzzing words.

Here, Fig. 2.2a represents the top five regions for cloud computing where India lead the search trend, Fig. 2.2b represents the top five regions for information security where Uganda lead the search trend, Fig. 2.2c represents the top five regions for Internet-of-Things and Fig. 2.2d represents the top five regions for predictive analytics, in both the cases Singapore leads the search trends.

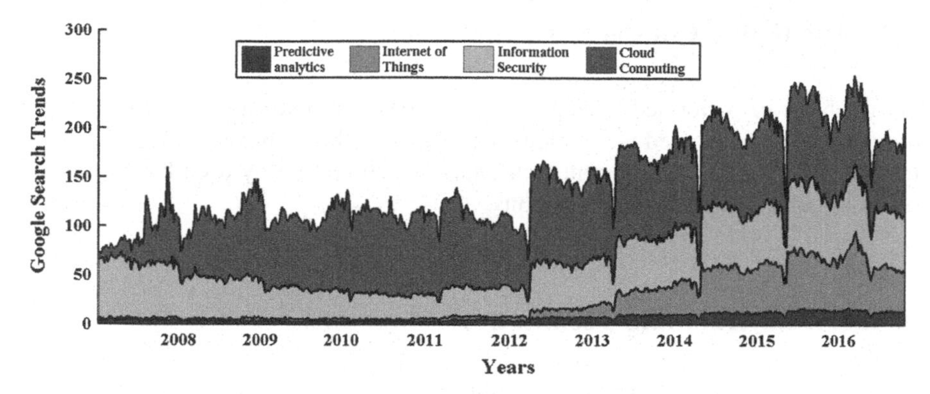

Fig. 2.1 Google search trends for buzzing words cloud computing, information security, Internet-of-Things, and predictive analytics

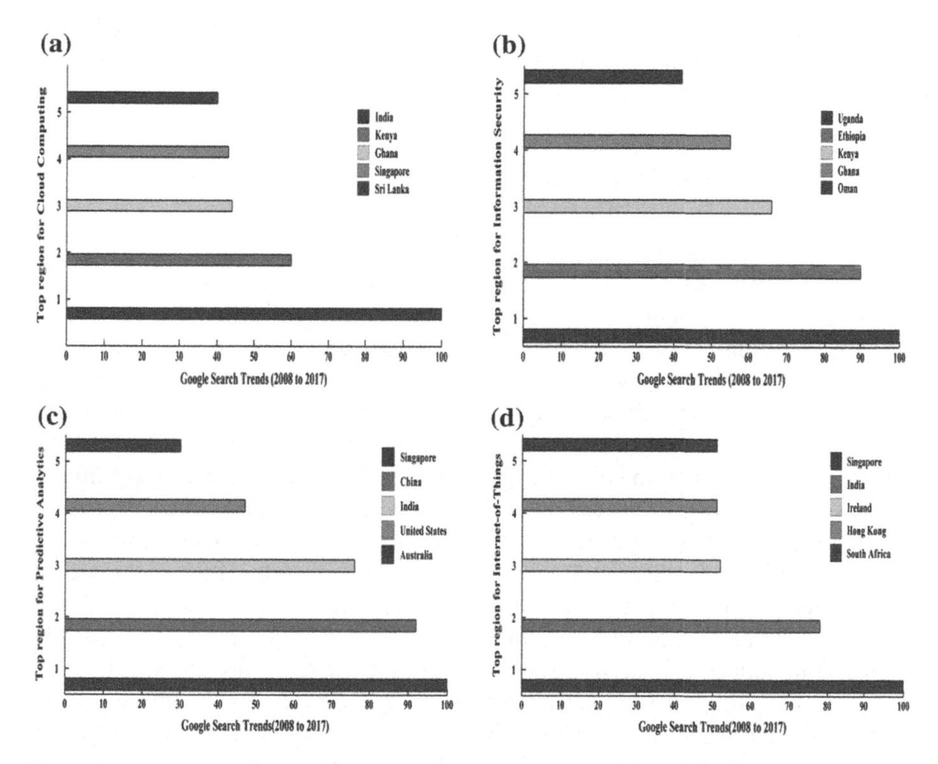

Fig. 2.2 Google search trends versus top 5 regions **a** cloud computing **b** information security **c** Internet-of-Things **d** predictive analytics

2.3 Predictive Computing Techniques

On the basis of performed operations and actions, the predictive computing techniques can be categorized into various categories. In this technical survey, we have tried to incorporate all the recent techniques which are mostly used by the developers to build their predictive solutions.

2.3.1 Data Handling Techniques

With the advancement in technology and computing techniques task of data generation and data processing has become easier. Various technologies like cloud computing, mobile computing, voice recognition, artificial intelligence and advanced application software are making prediction modelling possible. The predictive models are created whenever data is used to train a predictive modelling technique [9]. Table 2.1 summarizes various data handling techniques proposed by various researchers to find patterns in obtained data to perform smart computation.

2.3.2 Sustainable Techniques

Sustainable techniques are used to handle the issue of power consumption of computer systems and its devices. Majority of computer systems and devices are left unattended while they are active where they consume less amount of power. Similarly, if we discuss the same scenario from data centre's point of view then the power by such standalone device becomes disastrous. By using various predictive models and sustainable techniques, we can make predictions about the future of the device and computer system or device can be switched to energy saving state. In [10], Zhu et al. have presented an energy efficient reliable data gathering scheme in wireless sensor network environment. The proposed scheme is based on *Reed–Solomon code* and its enhanced version has been presented with intra-segment and inter-segment coding schemes. The authors have initially defined the optimization problem to derive the proposed energy efficient and reliable packet delivery scheme. In their obtained results, authors have claimed that proposed scheme is applicable in collecting data from sensor nodes at low-rate and low-power. In [11], Khan et al. have proposed a localization scheme *StreetLoc* for energy efficient smart phones using participatory sensing and have focused on the issue of data collection by these participatory nodes from urban streets. Authors have introduced the coverage metrics for the proposed full coverage, partial coverage and k-coverage schemes for the collected data from a street segment of the city. In their obtained results, authors have shown that proposed schemes can save a significant amount of energy. In [12], Abdullah and Yang have considered the issue of energy

Table 2.1 Data handling techniques for predictive analysis

S. No.	Author(s)	Proposed/Applied technique	Technical description	Parameters observed	Limitations and challenges
1	Wang et al. [119]	TVAS	Solved the issue of selection of destination sink node in the sub regions of sensor nodes and switching of aggregation scheme between TVAS and NS	– Time interval – Nearest neighbouring node, and – Data pressure	– Data pressure of sink node depends on threshold value – Metrics used for comparison average hop count, and network life time, and – Complicated concentration model
2	Villari et al. [120]	AllJoyn system	Proposed scalable solution to integrate system with lambda architecture [120] and processing is done using MongoDB	– Uses D-Bus to develop object oriented software independently – Patterns like regular, event based, and automated	– Difficult to handle complex scenarios – No connectivity support for inter domain – Large-scale smart environments management, and – Big data storage
3	Zhu et al. [121]	Permission-based RFID data collection algorithm (PRDC)	Proposed specification language can describe the complex tasks and algorithm is scalable during RFID data collection	– RFID collision – Temporal relations	– Communication overhead – Resource access in large-scale RFID systems
4	Wang et al. [122]	Data collection based on trajectory prediction (DCTP)	– DCTP designed for smart environments – Focused on reduction of incoming data and distributed prediction in real time	– Hidden Markov model – Message (ME) – Prediction of distance (PWD) – Prediction of trigger time (PTT)	– Churning of people is not considered, and – Walking speed is considered as constant

(continued)

Table 2.1 (continued)

S. No.	Author(s)	Proposed/Applied technique	Technical description	Parameters observed	Limitations and challenges
5	Zhu et al. [10]	Enhanced Reed–Solomon (E-RS) code	– Presented inter and intra-segment coding scheme	– Overall energy consumption	– Communication overhead – Lack of security
6	Sharma and Singh [123]	Seed block algorithm (SBA)	– Smart connectivity with remote client in absence of network and remote data backup	– Exclusive OR operation – Seed blocks	– More memory requirement for remote server – Huge difference in processing time
7	Xu et al. [45]	– Semantic data model – Data accessing method (UDA-IoT)	– IoT-based emergency medical services system is designed – Metadata model for ubiquitous data accessing, and – Building real-time system	– Value, annotation, and ontology, and – Entity oriented resource	– Processing cost of data – Coordination of multiple resources, and – Data security issue is not addressed

conservation in IoT and proposed a *message scheduling algorithm* to improve the efficiency of the system. The authors have also handled the faulty and failed node with the proposed algorithm. They have considered two main issues for energy conservation known as saving energy in battery powered objects and quick response to the query and the obtained results show the efficiency and effectiveness in service response and energy consumption. In [13], Brienza et al. have proposed the energy management system E-Net-Manager for various networked computers. They have proposed unique methods to reduce the energy consumption by using the soft sensors of computer systems like keyboard, mouse, bluetooth, Google calendar and PC activity soft sensors. In their results, significant energy saving has been achieved for short idle period. In [14, 15], Gupta and Singh have proposed a novel sustainable algorithm for prediction of CPU workload for minimizing the power consumption by personal computers. They have proposed a prediction model [16, 17] and algorithms for switching the current state of running computer system into power saving state.

2.3.3 Navigation Techniques

One major use of predictive computing is in navigation of vehicles or e-Transportaion. With the growth of technology, it has become possible to predict the shortest route for vehicle navigation. One can predict the navigation paths with lesser traffic rather than considering the busiest road networks in the city. These algorithms usually require a constraint to be placed on the system for effective prediction to take place. A Markov process is a stochastic process in which one can make predictions about the future state of the process based only its current state. Vehicle navigation paths are usually repetitive in nature due to natural constraints that limit the freedom of the driver. One of the most common natural constraints is time where most drivers just attempt to reduce the amount of time spent travelling between their origin and destination. The number of methods has been suggested for the prediction of vehicle navigation paths. Barth and Karbassi [18] have used a hierarchical tree data structure to perform real-time prediction of the navigation path that a vehicle may take for direct trips (source to destination). Their algorithm is recomputed as new data from the vehicle arrives while the vehicle is already in transit. Froehlich and Krumm [19] have discussed an alternative method where details of vehicle navigation path are collected and grouped by similarity. Each specific navigation path is assigned an index and stored. As the vehicle begins journey, the navigation path progresses and the algorithm attempts to match the current navigation path with an existing one. Although this allows for an initial prediction of the navigation path, the prediction is continuously updated as the journey advances. Kansal et al. [20] have discussed a sensor network for tracking using mobile phone devices. They have mentioned the fact that the prevalence of mobile devices and the increased availably of GPS technology makes them ideal nodes in a sensory network that focuses on the same GSM signals used for voice communication. Trials for autonomous vehicle navigations are on the way and soon one can expect the market release of such cars [21, 22]. Technologies like GPS, RFID [23], sensor networks and IoT have made predictive navigation approaches easier and implementable in all the sectors like automobile, aviation [24] and marine. Cao et al. [25] have used the GPS technology for constructing the minimum dominating set of navigation paths and used this data for selecting the best possible navigation path to drive. The authors have implemented the algorithms like marking process to find the dominating vehicle, Updating process to keep the data updated of neighbour's node, and cutting process to cut down the redundant vehicle information from the database. In [26], Davidson has presented three different algorithms that can be integrated into personal navigation systems. First, algorithm computes positioning for a map aided navigation system designed for land vehicles travelling on road network; second, algorithm is aimed at map aided vehicle navigation indoors and the third algorithm computes solution for the pedestrian navigation system. In [6], Pattanaik et al. have presented a smart congestion avoidance technique by estimating the scope of real-time traffic congestion on urban road networks and predicts an alternate shortest route to the destination. This technique utilizes the k-means clustering approach to estimate the magnitude of congestion and applies Dijkstra's algorithm to predict the shortest route. In [27], Su et al. have designed a shortest path computing algorithm for navigation of large commercial vehicles. They have integrated spatial data with the proposed algorithm. Authors have listed various characteristics of commercial vehicles types like bus, truck, trailer and passenger car, etc. Mitton and Rivano [28] have deployed sensors on bicycles to analyze the various road conditions for medical purpose and have gathered the data in real time from the deployed bicycles. The proposed system is in its preliminary stage and a lot of work is still remaining to obtain the results. In [29], Kranz et al. have discussed the concept of embedded interactions of objects in the day-to-day utilities. They have monitored these interactions using IoT. They have also listed various challenges for embedded IoT like Invisibility dilemma, implicit versus explicit interactions, context dependence, etc. They have presented the vision beyond ubiquitous computing for day-to-day computing using IoT. Similarly, in [30–35], the authors have presented the architecture and smart navigation techniques for vehicles and the presented techniques utilize the IoT and sensor networks for navigation of vehicle. Some of the major predictive navigational techniques are shown in Table 2.2.

2.3.4 *Intelligent Agents*

The vision of smart environment is possible because of growth in computing techniques and use of predictive models. The smart environments utilize the concept of prediction and communication among the various existing objects. This communication involves various intelligent agents at middleware level for remote access and control of information in smart environment. Chen et al. [23] have

Table 2.2 Predictive navigation techniques and their advantages and challenges

S. No.	Navigation algorithm	Purpose	Advantages	Challenges
1	Dijkstra's [124, 125]	To find shortest path with minimum cost	Can find the shortest routes or the routes with the shortest travel times between the origin and the destination	– Considering large number of nodes, and – Achieving better run time
2	Restricted search algorithm [124, 126]		Unlike search the entire circle as in Dijkstra, restricted search with the small area of the remaining part of rectangle	– Considering large number of nodes
3	*k*-top [127, 128]		Algorithm not only finds the shortest path, but also $k-1$ other paths in non-decreasing order of cost. k is the number of shortest paths to find	– Without considering the loops
4	A* search [124, 129]		Instead of finding the next node with the least cost, the selection of node is based on the cost from the start node plus an estimate of proximity to the destination	– Without improving worst case time complexity
5	Bellman-Ford [130]	To find shortest paths from a single source vertex to all of the other vertices	– Provides more efficiency in terms of nodes covered or path travelled – Edge weights in weighted digraph can be negative number	– Issue of scaling as the network topology changes
	A*-ants [131]	To find the best optimized multi-parameter direction between two desired points using electronic maps	A* algorithm is run before ants algorithm and updates (increases) pheromones of its resulted paths in ants algorithm	
6	CARPA [132]	Multipurpose route recommendation algorithm	It includes data preprocessing, accessibility metrics definition, and multi-objective path finding	– Without considering stairs, high curbs and busy intersections of roads
7	Kalman filter-based algorithm [133–135]	To support the navigational function of a real-time vehicle performance and emissions monitoring	Kalman filter-based algorithm generates pose estimation (position and orientation) information, which enables faster and more robust tracking	– Problems associated with Kalman filters is how to assign suitable statistical properties to both the dynamic and the observational models

proposed code centric RFID system, which is used to store the information based on smart agents and provide the instructions to system whenever action has to be performed. Still proposed system does not meet the goals like knowledge representation and situation-aware code interpretation. In [36], Huang et al. have done body posture analysis by using collaborative accelerometer sensors on different parts of human body like on neck, wrist, waist and thighs. They have tried to predict the odd situation with the help of predictive body posture analysis whether there is any accident like scenario or not. They are unable to determine the condition of the body once the accident has taken place and planning to use gyroscope for more accuracy of results for falling like scenarios. Jeong et al. [37], have designed and implemented large-scale middleware for ubiquitous sensor networks. This ubiquitous sensor network supports intelligent event processing, various types of sensors, real-time sensed data and management of collected metadata. Further, the authors are keen to extend the support for a variety of wireless communication protocols like ZigBee, Bluetooth, Code division multiple access, etc. In [38], Taylor et al. have focused on the distribution of electricity in future and proposed cost effective and intelligent solutions for electricity distribution network and management scenarios. In [39], Gaoan and Zhenmin have proposed an intelligent method for measuring the heart rate using mobile acceleration sensor. They have suggested using their proposed real-time heart-tracking algorithm to develop low-cost heart rate monitor device. In [13], Brienza et al. have focused on the issue of energy consumption by ICT devices and suggested some intelligent soft sensor methods like bluetooth based, Google calendar based, activity soft sensors, etc., to reduce the energy consumption using these devices.

2.3.5 Smart Objects Based Computing

With the growth of mobile networks, researchers and developers are targeting mobile devices as a smart object to perform predictive computing over collected data. These mobile devices are also working as a middleware for transferring data from one sensor node to another. In [40], Kortuem et al. have presented the concept of 'Internet of smart objects' and classified smart objects into three different categories (a) activity aware objects, (b) process aware objects and (c) policy aware objects. These smart objects are equipped with embedded display and buttons on it. They have categorized between these objects on the basis of parameters like awareness, representation, interaction and augmentation. Figure 2.3 represents various smart objects that can be connected with mobile like smart devices in a ubiquitous network. In this section, we have summarized various mobile computing techniques for ubiquitous computing of received data as shown in Table 2.3.

Fig. 2.3 Smart objects in a ubiquitous network [3]

2.4 Predictive Computing Frameworks

In [9], Kalechofsky has presented a simple framework for developing predictive and statistical models for modern business. The author has mentioned that predictive analytics includes various techniques like predictive modelling, machine learning and data modelling to make predictions. The predictive analytics, in turn, uses the predictive modelling framework to perform predictive computing. This computing technique can be used in almost every sector like healthcare, travel, marketing, e-commerce, etc., by using their modelling framework.

2.4.1 Healthcare Frameworks

The predictive computing plays an important role in the prediction of health related issues and sending the early alerts to the patients. These healthcare frameworks use various predictive modelling frameworks for predictions and that's why their accuracy may differ from one framework to another. Predictive modelling is widely in use in clinical research and analysis. In [41], Ng et al. have described a scalable predictive platform known as PARAMO (PARAllel predictive MOdelling) that can

Table 2.3 Smart object based various computing techniques

S. No.	Author(s)	Proposed technique	Application	Advantages	Challenges
1	Ko et al. [136]	Middleware architecture		Solves the issue of interoperability between sensor and middleware and provides intelligent service	Compatibility of smart object with sensors and service providers
2	Solanas et al. [137]	Smart health (s-Health)	Concept of smart health lies in between mobile health and smart cities	Data is collected from patients and also from sensing infrastructure of smart cities	Requires interactions between actors like government, physicians, researchers, and practitioners, etc. Privacy and security is a major challenge
3	Bottazzi et al. [138]	Socially aware and mobile architecture (SAMOA)	Framework allows creating roaming social networks for mobile users anytime anywhere	Middleware solution to address social network management details	Lack of security and performance issues
4	Soliman et al. [139]	Smart home application	Integrated IoT, cloud computing, and Zigbee technology to design the application	Easy to use with improved efficiency of data access	No focus on security and privacy like issues and interoperability with other protocols and smart objects
5	Kortuem et al. [40]	Interaction between smart objects	Categorized smart objects in three categories	Included interaction and social aspects in design of smart objects	– Developed workflow models and ad hoc combination techniques, – Implementation of these techniques is still remaining
6	Mitton and Rivano [28]	City bike network	Deployed sensors on city bikes while cycling to gather data related to road, environment, or medical purpose	– New concept to promote eco smart driving using IoT – Basis model is proposed for city bike network equipped with sensors and base stations to collect the data	– Real-time implementation is complex in nature and involves more base stations as bicycle is not a stationary object and deployed sensors on bike have range limitation – For transferring of data with one base station object can't wait at one place
7	Siebert et al. [140]	LASEC algorithm and novel mechanism A-Ack	Collaborative composition approach has been implemented by the algorithm between service providers and A-Ack mechanism restricts the number or exchange messages for each service	– Composition algorithm is designed in such way to minimize the cost while object is not stationary – Network overhead is reduced by minimizing the number of messages to be exchanged	– Two layer model i.e. communication layer and request layer – Continuous energy consumption by nodes while they are in composition with other nodes – Object must be in the range of nodes

be used by many predictive applications. Brooks et al. [42] have stated that healthcare organizations can implement a predictive model that uses business intelligence (BI) to improve clinical efficiency. They have proposed a framework for developing a domain specific BI model for an expert practitioner in the field. In [7], Gupta et al. have proposed an IoT-based cloud centric health framework to monitor user's activities in sustainable health centres. This framework regularly predicts the user's activities and stores the details of various parameters like heart rate, pulse rate, timing of activity session, etc., in the cloud database and sends alerts to health care professionals of the user wherever emergency like scenario arises.

In [43], Lu et al. have presented the SPOC framework for secure and privacy preserving computing in mobile health care. This framework extends its functionality in the pervasive environment and introduces the user-centric privacy controls. In [44], Zhang et al. have presented three tier system architecture for healthcare system based on ubiquitous sensing. Tier-1 consists of various objects equipped with ubiquitously distributed sensor nodes. Tier-2 consists of ubiquitous technologies like wireless sensor networks, internet, WiMAX, 3G/4G, etc., to transfer the sensed data to doctors or medical practitioners. Whereas, Tier-3 is an information processing centre and consists of nodes like medical practitioners, doctors, data centres, servers for processing of collected data related to patient's health and to provide the necessary action to be taken by the patient. Similarly, Xu et al. [45], have proposed the semantic data model for interpreting the IoT data and also a data access mechanism has been proposed for emergency medical services in ubiquitous environment. Hu et al. [46], have also used the IoT network of physical objects for monitoring and predicting the health condition of elder person living at home. They have proposed an intelligent and secure health monitoring framework for finding the elder's activity and elders can use the mobile device for getting connected with the IoT network. In [47], Zhao et al. have presented the predictive model for finding the effect of multiple sclerosis in patients. They have used logistic regression and machine learning techniques in predicting disease course. In [48], Abreu et al. have studied about the recurrence of breast cancer and presented a predictive model based on combination of machine learning techniques for accurate prediction of recurrence events. In [49], Rana et al. have introduced the concept of changing interventions on different datasets and proposed a predictive framework that models interventions explicitly. In [50], Sakr and Elgammal have discussed some major challenges in healthcare systems which can be resolved using modern technologies like cloud computing, IoT and big data. The aspects of these technologies can be different that vary from communication to data storage. Authors have proposed a framework known as '*SmartHealth*' and described its various applications in the healthcare domain. Ifrim et al. [51], have highlighted the current status of the evolution, trends and research on IoT from e-Health perspective and performed a comprehensive survey.

2.4.2 Smart Home Frameworks

The potential technology used for designing the smart home varies from simple sensors to detect the position of door to more sophisticated sensor systems where the sensors are equipped into various objects and humans reside inside the house. Mulvenna et al. [52] have examined the role of context aware computing in smart home environment and monitored the activity of each individual in the house. Authors have developed various frameworks for monitoring the activity. These smart objects equipped with the sensor, collects the data about user's offline activity performed at home and continuously transmit it on the Internet. Apthorpe et al. [53] have mentioned that IoT devices have always-on sensors that capture the data constantly about the user's physical environment. They have developed a prediction strategy from the obtained data for passive network observer and to find about the various possibilities related to the user. In [54], Raj has presented a framework for smart monitoring of home and its security in owners absence. He has designed the context aware protocol for this system. This system is secure, reliable and user friendly and combination of Zigbee, Wi-Fi and body area network like technologies. By making the use of pervasive computing this framework is known as smart home system where user is connected with the various devices in home remotely. In [55], Aquino-Santos et al. have also presented the use of ubiquitous computing in developing smart home applications and interoperability of these applications. They have discussed the challenges like isolation of subsystem by executing several instances using hypervisor for the connected devices, and implementing a micro middleware. In [56], Hong et al. have analyzed the user's habits and behaviour in a living space and discussed the context aware model for smart home systems and the proposed model is shown in Fig. 2.4.

This model connects the various household devices with the user behaviour and process the collected information in an intelligent environment.

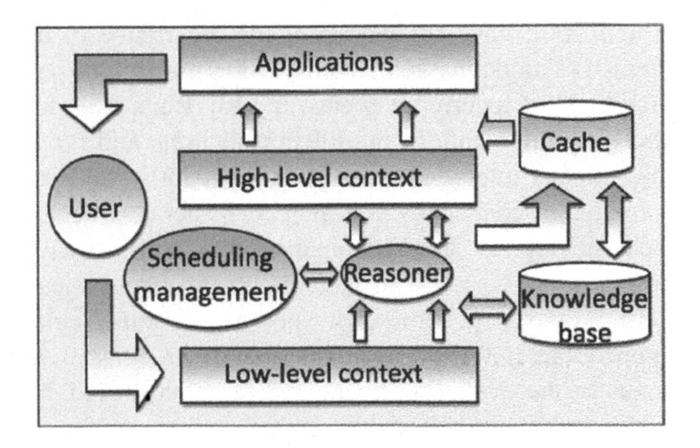

Fig. 2.4 Context aware smart home system model [56]

2.4.3 Navigation Framework

Currently, the road transport of goods or passengers relies on tracking technology. In designing of smart navigation systems, GPS data may be augmented with Wi-Fi and GSM signals to be used to provide location information of vehicle transporting the goods and passengers [57]. At present these systems suffer from limitations like reduced reliability in areas that are not permeated by the necessary GSM or Wi-Fi signals, or areas in which the GPS satellites do not have sufficient coverage. In [5], Malekian et al. have designed and implemented a system which is capable of predicting the navigation path of a vehicle on the basis of the existence of driver's existing driving practices. This system is capable of using RFID based information about navigation paths, in conjunction with predictive algorithms based on the Hidden Markov Model (HMM) to accurately determine the vehicle navigation paths in advance. In HMM, the sequence of observed output values provides information about the sequence of states. If modelled using a HMM, then the observer will only observe a sequence of output tokens directly. Baum et al. [58], have described a model based on this information. The observer can then attempt to infer the sequence of states that yielded the observed output sequence. HMM is an acknowledged tool for predictive solutions to systems that can be modelled as Markov processes. In [59], Simmons et al. have proposed the usage of the HMM to perform predictions on a vehicle's navigation path. In their method, the historical driver data is gathered using GPS information. This is used to supply parameters to the Hidden Markov model. They were able to achieve results of above 98% accuracy in most cases, although the navigation paths they tested had very few places in which choices were required. In [60], Herbert et al. have a proposed a modular framework FaSTrack that provides a safety controller and can be used with most current paths. In [61], Jabbarpour et al. have presented the general framework for vehicle traffic routing system, known as VTRS and used for mitigating traffic congestion on roads. VTRS gathers traffic-related data such as vehicles' speed, travel time and density, user preferences and alternatives paths for preparing the routing tables by calculating the fitness function. In [62], Yang et al. have proposed an autonomic navigation system (ANS) operating over vehicular ad hoc networks (VANETs) for predicting the future vehicle density and adopts hierarchical algorithms for route planning and time dependent routing algorithm for traffic prediction as shown in Fig. 2.5.

In [63], Cebecauer et al. have presented a framework for integrated real-time network travel time prediction of vehicle based on probe data. This framework is capable of predicting short-term traffic conditions, real-time vehicle routing and for trip planning. They have used hybrid probabilistic principal component analysis (PPCA) methodology for short-term prediction of vehicle path. Similarly, a number of frameworks and approaches have been proposed for finding the predictive trajectory guidance, prediction of next turn at road junction and motion planning for autonomous vehicle navigation [64–67].

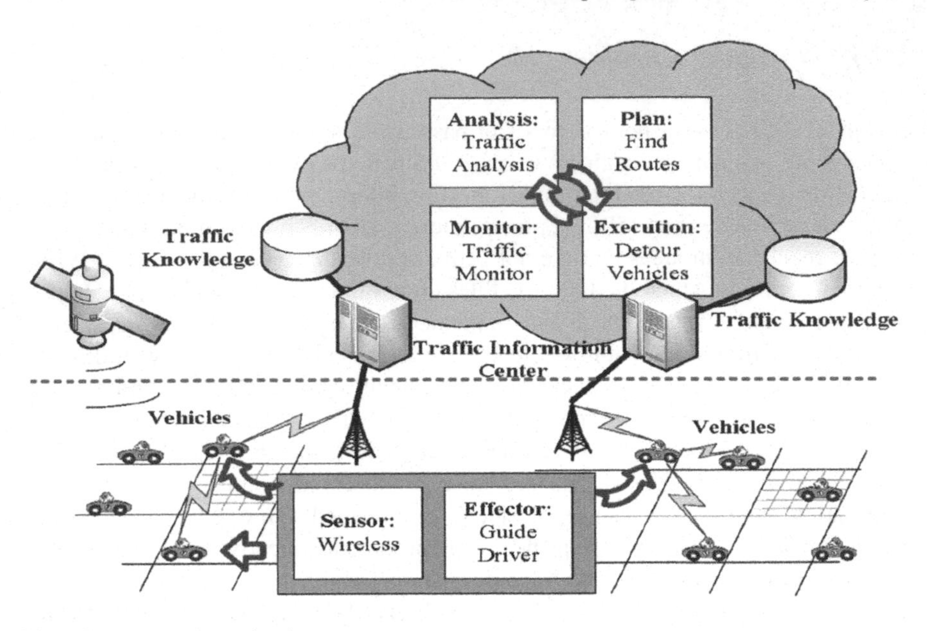

Fig. 2.5 Autonomic navigation system

2.4.4 e-Commerce Framework

Since last decade, a lot of significant developments have taken place in e-commerce market and its applications. The organizations have completely changed the way of marketing by implementing the various data mining rules for predicting the price of product, behaviour of consumer and selecting the appropriate marketing strategy, granting them to make intense and knowledge-driven results. Similarly, change in marketing models is taking place from traditional website based system to knowledge-based system to recommender system. A recommender system predicts about the customer's choice by exploiting ratings made by the customer for the same or similar product in the past. In [68], Qiu has proposed a predictive model, COREL to make purchase behaviour prediction of customers. He has investigated the three factors that significantly affect the purchasing decision of customer in online shopping that is the needs of customers, the popularity of products and the preference of the customers. He has explored associations between products, exploiting them to predict customer needs. In [69], Gupta and Pathak have focused on the dynamic pricing of the product where prices vary according to market demand of the product. Nowadays, dynamic pricing concept is being used by almost all e-commerce sites including retail, automobile, tours and travels, grocery stores and a lot many. They have proposed a model for determining the purchase behaviour and pricing strategy for online customers. This proposed model consists of stages like Data Collection, Preprocessing, Attribute selection, Grouping of customers, Dynamic pricing and Predictive analysis as shown in Fig. 2.6.

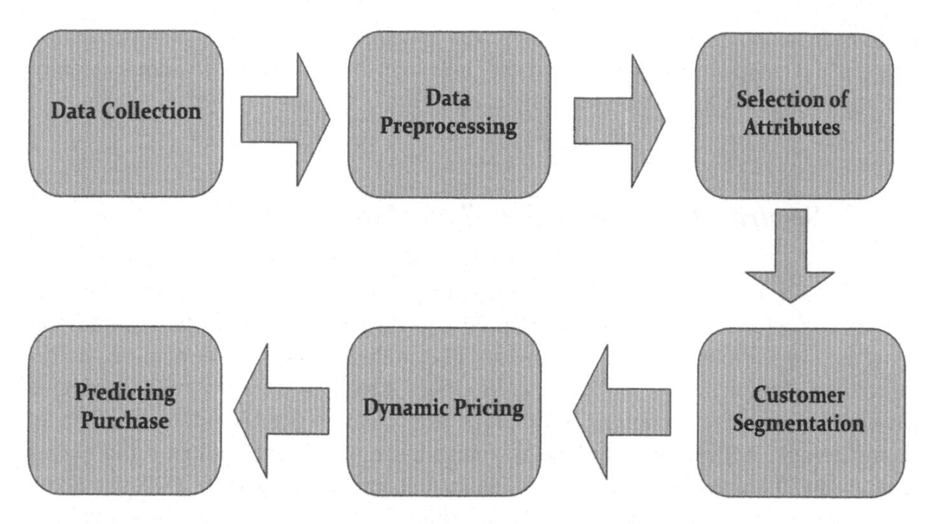

Fig. 2.6 Framework for predicting online purchase based on dynamic pricing

This framework also includes various techniques of Machine learning, Data mining and Statistical methods for predicting the online purchase behaviour of customers by selecting an appropriate price range. In [70], Ahmadi has proposed a framework, e-CLV (Electronic customer lifetime value) for predicting the online customer's behaviour. Proposed model considers real option analysis to predict all future states with probability of each of them. In [71], Lo et al. have tried to characterize, understand the customer's behaviour for developing predictive models of user purchasing intent. They have identified some set of general principles and performed large-scale longitudinal study to model user purchasing intent across many discovery applications. Authors have classified the user's action into four classes known as (i) *searching for a content* (ii) *exploring contents by using provided links* (iii) *getting closer with particular piece of content and* (iv) *saving contents to retrieve it later*. In their predictive analysis, authors have tried to find how engagement in these actions predicts users' future purchasing intent or activity. Similarly, in other studies [72, 73] related to prediction of consumer behaviour, authors have used different perspective of analysis like Badea [72] have used the concept of artificial neural network for predicting consumer behaviour, and Naumzik et al. [73] have performed image sentiment analysis on their proposed model for predicting the increase in rent prices.

2.5 Information Security Techniques

'Data' is becoming a vital component for all kind of computing techniques. Several issues arise with the handling of data like data storage, data access and data usage issues, etc. As discussed earlier, over the period of time various security techniques have been proposed by the researchers and developers to establish confidentiality, privacy and trust while handling of data. Here we have summarized various security techniques related to cloud computing and IoT.

2.5.1 Security Techniques for Cloud Computing

Security is a major issue in cloud computing as it leads to various types of vulnerabilities while handling of data. Cloud storage and data has to be secured and must ensure various security parameters like authentication, authorization, confidentiality, integrity, availability, etc. In [74], Arockiam and Monikandan have proposed a security technique AROcrypt that ensures confidentiality of data in cloud storage. AROcrypt is one of the symmetric encryption techniques and makes use of ASCII values for processing of plain text into cipher text. In [75], Alsulami et al. have investigated many security techniques and models for cloud computing to main the data integrity and confidentiality. They have identified few of the techniques like encryption, anonymization, separator and multilayering that can have an effect on data integrity and confidentiality. Among these techniques, encryption technique is most widely used for cloud security. In [76], Zhou et al. have presented a scheme to control and prevent unauthorized access to data stored in the cloud. They have proposed a role-based encryption (RBE) technique that integrates the cryptographic technique with role-based access control (RBAC) model. They have also proposed a RBE-based architecture for secure data storage in public cloud. To maintain the privacy of data, policy based encrypted data access approach has been used in which users who satisfy the access policies can decrypt the data using their private key. In [77], another work, Bokefode et al. have used encryption technique AES and RSA for encryption and decryption of data and RBAC used to provide control to users as per their role. In [78], Gugnani et al. have focused on to provide confidentiality to the user while using cloud-based web services, and proposed an approach for selective encryption of XML elements so as to provide confidentiality and to prevent XML document form improper information disclosure. Figure 2.7 represents XML DNA Encryption/Decryption to embed confidentiality. Authors have considered the XPath Injection attacks which take place when web site uses user-supplied inputs to form XPath query for XML data. Terec et al. [79], have discussed the implementation of various cryptographic techniques in Java, MATLAB and Bio-Java and also represented how DNA encryption is implemented in three of them. Besides XML encryption, we have SSL and XML signatures to secure the internet transmission. Users interact with the cloud using XML files and then these XML files and their contents need to be protected so as to safely transfer the confidential information. Also, the communication among Web Service and the clients are mainly done through plain-text XML formats like SOAP messages and WSDL [80].

In [81], Dinesha and Rao have proposed a secure cloud transmission protocol (SecCTP) to maintain data integrity, confidentiality, access management and

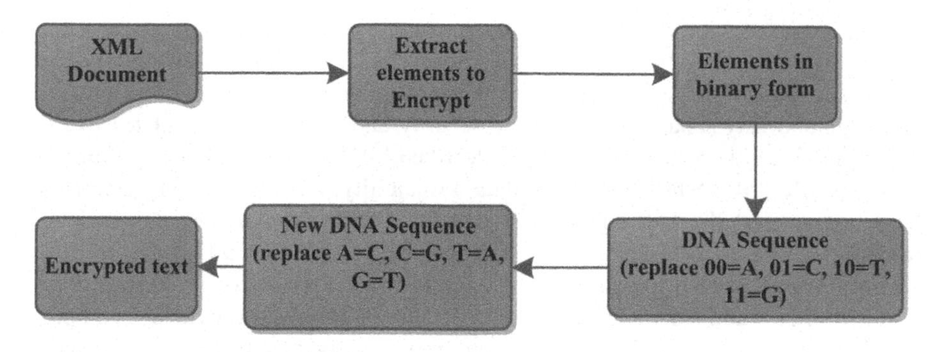

Fig. 2.7 XML DNA encryption approach

identity in the cloud. Proposed technique SecCTP deploys Multi-Dimensional Password (MDP) system, Multilevel Authentication scheme (MLA) and Multilevel cryptography (MLC) system to handle various cloud security issues of authentication, identity and confidentiality. In [82], Yu et al. have proposed an efficient and practical identity based Remote data integrity checking (RDIC) protocol which is based on key-homomorphic cryptographic schemes and maintains perfect data privacy. In perfect data privacy, protocol leaks no information of the stored files to the verifier. In [83], Zhou et al. have stated that the integration of IoT techniques with cloud computing leads to new challenging security and privacy related threats. They have proposed unique security and privacy requirements for cloud-based IoT and addressed the challenging issues of secure packet forwarding and efficient privacy preserving authentication by proposing a new efficient privacy preserving data aggregation without public key-homomorphic encryption. In [84], Choi and Lee have focused on various security issues related to the public sector that restrict any organization from the use of cloud platform. They have proposed a methodology for information security management and adopted a Delphi approach to establish the classification criteria in objective and systematic manner. In [85], Gaetani et al. have focused on data integrity issue in cloud computing and used a *Blockchain-based* method for handling these issues. Blockchain has recently emerged as a fascinating technology that consists of consecutive chained blocks containing records that are replicated on the nodes of a P2P network. These records witness transactions occurred between pseudonyms. They have proposed an innovative blockchain-based database that permits balancing strong integrity guarantees with appropriate performance and stability properties.

2.5.2 *Security Techniques for Internet-of-Things*

The architecture of IoT converts living and non-living entities into smart objects that can be monitored or used continuously by using internet technologies. These smart objects can communicate with each other, and can respond to the changes in surroundings intelligently. This intelligent feature derived from IoT architecture includes new security risks and privacy issues. So, maintaining user security and privacy is one of the crucial issues and need to be addressed thoroughly. In [86], Dabbagh and Rayes have presented the security and privacy issues related to IoT platforms. Various security challenges identified by them are Scalability, Multiple verticals, Multiple technologies, Big data, availability, remote locations, etc. They have categorized the IoT architecture into three different domains that include: (a) IoT Sensing Domain, (b) IoT Fog Domain and (c) IoT Cloud Domain. Various security attacks related to cloud domain and IoT are summarized in Table 2.4.

In [87], Nia and Jha have presented a detailed survey and provided a comprehensive list of vulnerabilities and countermeasures against them. Authors have discussed three widely known IoT reference model, i.e. (a) Three-level model [88], (b) Five-level model [89] and (c) Cisco's Seven-level model [90] and discussed various possible applications of IoT. In [91], Singh et al. have discussed state of the art of various lightweight cryptographic primitives that consists of lightweight block ciphers, hash functions, stream ciphers, high-performance systems and low resources devices for IoT environment. They have also analyzed several lightweight cryptographic algorithms like AES, DES, Twine, Seed, RC5, PRESENT, etc., on the basis of their key size, block size, a number of round parameters.

2.6 Information Security-Based Frameworks

As discussed in the previous section, various information security techniques have been evolved to mitigate or to reduce the vulnerabilities in an existing software system, used by business organizations or its customers. For effective implementation of information security and privacy techniques in the system developers and software designers need to modify the existing architecture or framework of software system. It is observed, that in most of the cases these architectures or frameworks are responsible for security-related issues. The reason for this is that security parameters are overlooked by the designers and developers while designing of system. In this section, we have discussed various information security frameworks related to cloud computing and IoT.

2.6.1 Cloud Computing-Based Security Frameworks

Rising demand for storage of data in a cloud environment by the business organizations and customers requires a lot of changes in existing architecture or framework of software systems. Current system design and deployment techniques must be capable enough to accommodate these changes into the system and must ensure the security and privacy of new cloud-based frameworks and architectures. In [92], Chang et al. have presented a multilayered security framework for business

Table 2.4 Security attacks in IoT-based cloud domain

S. No.	Security attacks	Security violation	Reason	Technique(s) to be considered to resolve
1	Hardware Trojans [141–143]	– Confidentiality – Integrity – Availability – Accountability – Auditability – Trustworthiness – Privacy	– Alteration of integrated circuit design – Activation of Trojans based on some external/internal factors	– Circuit design modification, and – Improving Trojan activation method
2	Physical attacks [144–146]		Shared access to components, or network resources	– Hard isolation of devices and software's – Circuit design modification
3	Spoofing/Tag cloning [147]		– Weak security mechanism – Easy to intercept, read and save messages	– Personal firewall – Cryptographic scheme – Hard isolation of devices and software's – Distance estimation – Kill/Sleep command
4	Malicious injection [148]		– Insufficient validation of the input	Pre-testing and validation of inputs
5	Non-standard framework and weak testing [149]		Non-standard coding flaws	– Pre-testing – Designing good framework and policies
6	Virtual machine migration (VM migration) [86]	– Confidentiality – Integrity – Availability	Software bugs, no permission for migration, malfunctioning of device	Server authentication
7	Hidden-channel attack [86]	– Confidentiality	Shared access to components, or network resources	– Hard isolation – Cache Flushing – Noisy data access time, and – Limiting cache switching rate
8	Theft-of-service attack [86]	– Availability	Periodic sampling	– Fine-grain sampling – Random sampling

(continued)

Table 2.4 (continued)

S. No.	Security attacks	Security violation	Reason	Technique(s) to be considered to resolve
9	Insider attack [86]	– Confidentiality – Integrity	Lack of trust	– Homomorphic encryption – Divide the data into multiple chunks and use secret key with certain permutations
10	DoS [150–152]	– Availability – Accountability – Auditability – Privacy	– Node outage because of battery draining – Sending undesired set of requests that seem to be Legitimate – Node outage because of unintended error, sleep deprivation, code injection, etc.	– Securing firmware – Personal firewall, and – Cryptographic scheme
11	Eavesdropping [153]	– Confidentiality – Privacy	– Weak security mechanism – Easy to intercept, read and save of messages	– Personal firewall – Cryptographic scheme – Hard isolation of devices and software's – Blocking
12	Routing attacks [154–156]	– Confidentiality – Integrity – Accountability – Privacy	– Changing routing information	Reliable routing
13	Integrity attacks [157, 158]	– Confidentiality – Integrity	– Manipulating the training dataset – Direct access of server or computing nodes for manipulation	Outlier detection

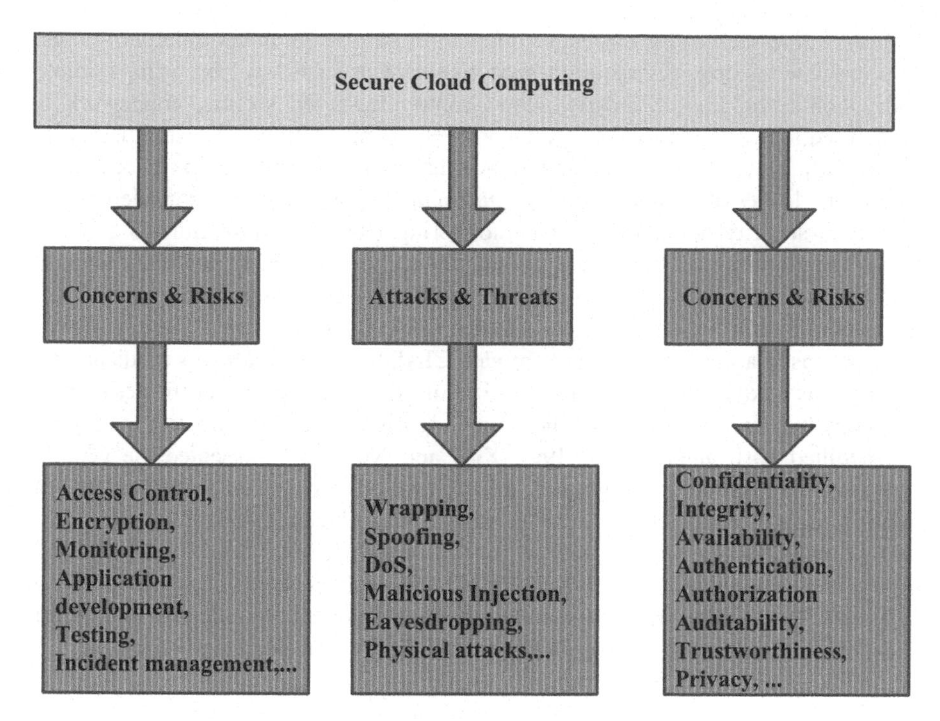

Fig. 2.8 Framework for secure cloud computing

clouds. This Cloud Computing Adoption Framework (CCAF) integrates three major security technologies known as firewall, identity management and encryption. In [93], Mushtaq et al. have presented the quad layered framework for data security, data privacy and data breaches. This layered architecture prevents the confidential information with a variety of quad security layers like Secure Transmission of Data, Encrypted Data and its processing, Database Secure Shell and Internal/external log Auditing. They have also presented a new auditing mechanism where customers can design their own rules for making auditing better. In [94], Youssef and Alageel proposed a framework to identify cloud specific security and privacy challenges, attacks and risks. They have also proposed a security model to perform generic cloud computing and to protect security and privacy requirements from vulnerabilities as given in Fig. 2.8.

In [95], Matte and Kumar have focused on storage of data, and stated that cloud supports data storage in distributed environment. They have provided the solution for security of data in multi-cloud. Multiple copies of data are stored in encrypted form on different clouds. Authors have used plain cipher algorithm for encryption purpose. In another work, Dorairaj and Kaliannan [96] have focused on issue of data migration in cloud, which is accessed by several users. Data protection from unauthorized access becomes important to both organizations and customers. To handle the issue of data sensitivity, authors have proposed an adaptive multilevel security framework that provides adequate level of security under different classes. This framework provides required access control at each level by using suitable encryption techniques. Similarly, cloud computing-based security framework is being used in every domain like healthcare, vehicle navigation, eLogistics, banking, etc. In [97], Ondiege et al. have mentioned that poor implementation of security in healthcare leaves the patients' data vulnerable to attackers and considered that providing security to remote patient monitoring (RPM) infrastructure is a major issue. They have proposed a new identification technique NFC in their new security framework for monitoring of remote patients in multi-user environment and to keep the patient's information secure. In another work, Jaganathan and Veerappan [98] have proposed a new cloud storage model, CIADS, to keep patient's medical data secure. They have implemented authorization service through certificates, confidentiality by implementing new encrypting algorithm and data integrity is ensured by modified hash algorithm. In [99], Xiao and Xiao have presented the generic ecosystem for cloud security and privacy which employ an attribute driven methodology. This ecosystem employs five security/privacy attributes: (a) Confidentiality, (b) Integrity, (c) Availability, (d) Accountability and (e) Privacy-preservability, as shown in Fig. 2.9. Authors have considered security and privacy separately and demonstrated the connection among vulnerability, threat and defense mechanism for the mentioned attributes in cloud environment.

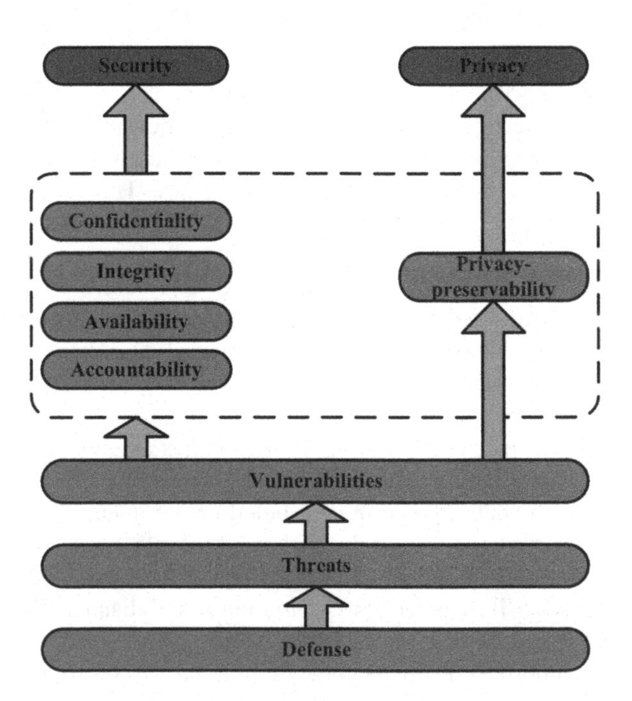

Fig. 2.9 Generic ecosystem based on attributes for cloud security and privacy

2.6.2 IoT-Based Security Frameworks

Significant advancements in the field of information and communication technology (ICT) and wireless technology, lead to development and adoption of various frameworks which are based on cloud computing, IoT, Big data, etc. It is predicted that by 2020, around 50 million of things will be connected to the Internet via IoT [100]. Usage of IoT-based frameworks in every business sector like healthcare, smart home, agriculture, logistics and transportation, is increasing very rapidly as it provides anyone, anytime, anyplace and anywhere type of frameworks for the all connected living and non-living things in the network [101]. However, these technological advancements and adopted frameworks also bring many issues. Security and privacy are considered at the top of all issues that need to handle intelligently for global adoption and use of IoT technology by the humans [102]. Park and Shin [103], have proposed a general security assessment framework for IoT services. They have applied integrated fuzzy multicriteria decision-making (MCDM) approach which uses an analytic network process (ANP) in combination with the decision-making trial and evaluation laboratory (DEMATEL) technique to increase the sensitivity of interrelationships among diverse security requirements. This framework is shown in Fig. 2.10.

In [104], Ge et al. have proposed a framework for modelling and assessing the security of the IoT and provide a formal definition of the proposed framework under five phases known as (a) *data processing*, (b) *security model generation*, (c) *security visualization*, (d) *security analysis and* (e) *model updates*. This framework identifies all possible attack paths in the IoT and evaluates the security level of the IoT through security metrics. In [4], Kang et al. have focused on human centric smart home services and proposed an enhanced security framework for smart

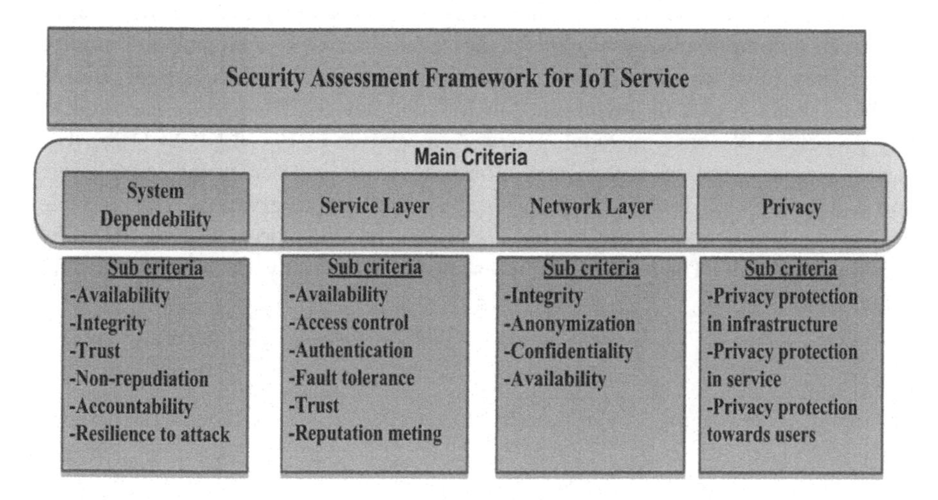

Fig. 2.10 Security assessment framework for IoT

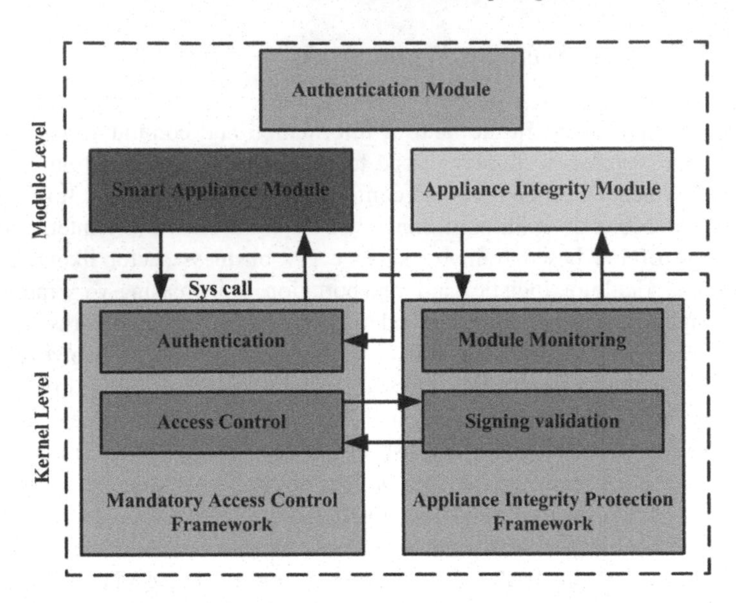

Fig. 2.11 Security framework for smart home

devices in a smart home environment. As shown in Fig. 2.11, this framework is made up of smart appliance module, appliance integrity module, mandatory access control framework and appliance integrity protection framework. This framework provides the security service for ensuring device authentication, integrity and availability.

In [105], Ngu et al. have focused on realization of middleware technologies in IoT systems which represents that software framework play as an intermediary between IoT devices and applications. They have designed an application for real-time prediction of blood alcohol content using smart-watch sensor data and presented a comprehensive survey on the capabilities of the existing IoT middleware. They have also presented a thorough analysis of the challenges and the enabling technologies in developing an IoT middleware. They have captured the key properties of some trusted IoT system as shown in Fig. 2.12. In [106], Ukil et al. have presented privacy preservation framework as a part of the IoT platform and a data masking tool for both privacy and utility preservation. This provides negotiation based architecture to find a solution for utility-privacy tradeoffs in IoT data management. They have also presented a case study on e-Health for this framework.

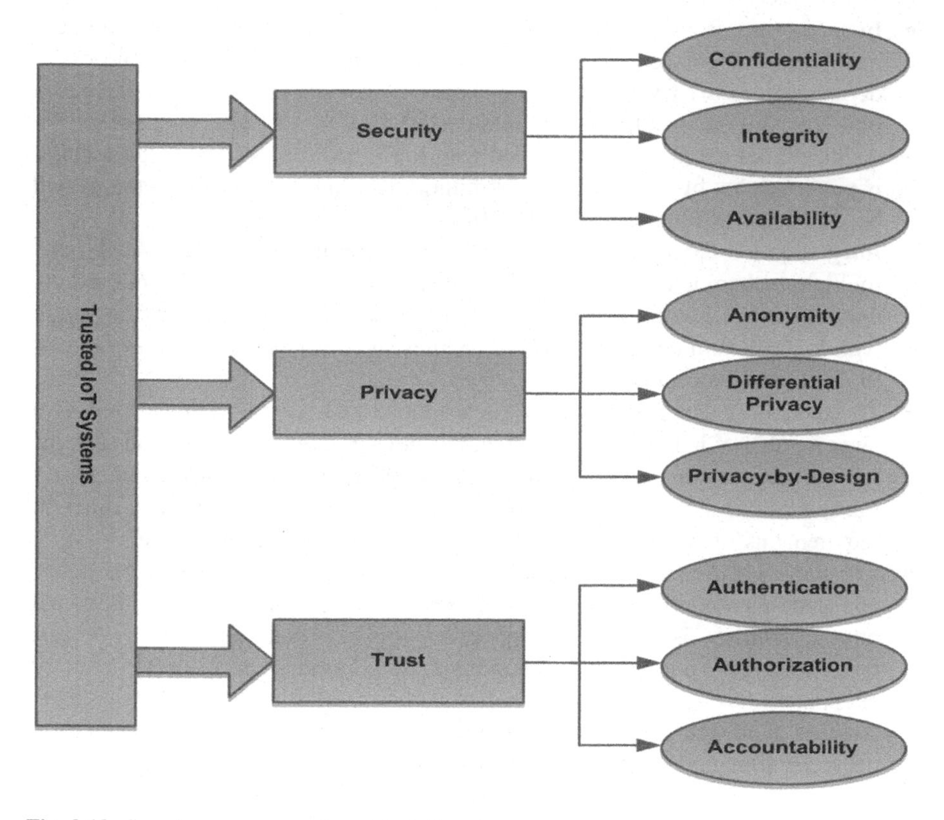

Fig. 2.12 Security, privacy, and trust in IoT systems

2.7 Challenges and Discussions

A primary focus of this chapter is to discuss various techniques and frameworks available for predictive computing and information security. We have found that the security issues in cloud computing and IoT will become a challenging task in near future. According to new research released by TRUSTe, 35% of online US consumers now own at least one smart device other than a smart phone, and the most popular devices are smart TVs (20%), in-car navigation systems (12%), followed by fitness bands (5%) and home alarm systems (4%). In this section, we have summarized few major challenges related to cloud computing and IoT which are listed as follows:

- As per findings from 2015 U.S. IoT Privacy Index, only 79% of consumers are concerned about the idea of their personal data being collected through smart devices, while 69% believed they should own any such data being collected [107].

- In supporting security, privacy and trust mechanisms within IoT middleware has been recognized as a critical and important issue for the successful deployment of IoT applications, and is deemed as one of the major challenges in both industry and academic communities [105].
- As the digital market will be flooded with IoT-based devices and cloud users. It is predicted that by 2020, around 50 million of living and non-living things will be connected to Internet via IoT [100].
- Integration of cloud computing, wireless sensor networks, RFID and other multiple technologies with IoT will increase the complexity of overall system exponentially, as each technology has its own vulnerabilities and integration with IoT will sum up all these vulnerabilities and introduce new security threats to devices and systems [86].
- The modelling security of the IoT is a difficult task as it is characterized by a large number of heterogeneous and mobile devices and facing numerous threats.
- IoT paradigm consists of several verticals that span over e-Health, intelligent transportation, agriculture, smart home, etc. All these verticals have different requirements of security and privacy at device level and at user level.
- IoT framework consists of a number of sensor nodes and scalability of IoT application stored in the cloud, with respect to utilization of CPU, memory and other resources is a challenging task. An IoT solution needs to scale cost-effectively, potentially to hundreds of thousands or even millions of endpoints.
- Another challenge from these endpoints is that they generate heavy amount of data over time and require some effective mechanism in cloud for security and privacy of this data. This large volume of data is also known as Big Data [86].
- The ownership of the data collected from these endpoints must be clearly established and the endpoints and reading devices from the IoT Things should each be equipped with privacy policies. This collected data is further processed to obtain useful information related to health analysis, vehicle route prediction, monitoring a home and environment, etc. [106].
- In [103], Park and Shin have considered service availability challenge at the top priority with increased attention on infrastructure security for networks and systems.
- Further, the nodes in IoT have limited energy and computational power on which implementing complex security measures, is a challenging task. To handle this challenge, security mechanism needs to be developed in context to IoT resources that focus on particular type of attacks like DoS, spoofing, etc., this will help to utilize less energy and computational time.
- In [87], Nia and Jha have also discussed about two emerging security challenges: (1) *Exponential increase in number of weak links*—Currently, available devices in the market don't support complex cryptography algorithms because of these restrictions and this led to number of weak links in the system that can be exploited by the attackers. (2) *Unexpected usage of data*—Growing use of IoT technologies has led to sensor-based connectivity in day-to-day living, this scenario leads to unexpected use of user's data collected by these sensors.

- In another approach, attack paths could be found with the help of framework if node vulnerability and network reachability information is given [103]. The attackers can access these IoT devices and resources with the help of cloud and can use these devices as zombies, so protection of each node becomes important.
- Most of the IoT-based devices are mobile in nature and this mobility has a great influence over security, because attack surfaces changes with the change of network. This challenge involves designing of mobility model for protection of IoT device node in hanging network environment.
- Designing IoT applications and services is yet another challenge as IoT middleware must be available in cloud and on the edges like IoT devices, gateways, etc.
- A big challenge is to ensure the security of IoT application and privacy of users along with semantic service discovery in which a failed IoT service gets replaced with available ones in the network without causing any disruption to the user [105].
- As reported, lack of security mechanisms, auditing mechanisms, data integrity and service level agreements are serious concerns in cloud computing. Some security models have been proposed related to proof of retrievability [108], anonymity based system [109], privacy stabilizing architecture [110], process of access control [111], preserving cloud computing privacy (PccP) model [112] and public auditing mechanism-Oruta [113, 114], to overcome mentioned challenges.
- According to NIST [115], various challenges associated with the cloud computing platform are: designing of policies, standards and procedures that are sufficient to defend organizations from threats.
- The distributing the roles and responsibilities among team members to implement security policy is yet another challenge that should be followed with effective planning at each stage of system's life cycle.
- Implementations of security policies are considered in ad hoc manner in cloud environment to satisfy some set of organizational needs to minimize the risk of threats. The future challenge would involve real-time measures to provide assurance against organizational goals.
- Business outsourcing is common phenomenon of organization and number of organization opt outsourcing to reduce the overall business complexity. The cloud computing facilitates organizations with computational and storage resources at reduced operational cost.
- In cloud environment, data owner lacks full control over outsourced data and finds its management untrustworthy as data may get exposed to various insider and outsider attacks or data leakage related issues could be there [100]. Therefore, maintaining a data confidentiality and privacy is a challenging task to gain users confidence for adopting cloud computing.
- Implementing proper access control mechanism could be challenging issue to maintain user's privacy and unauthorized access to data. As mentioned, in cloud computing data is stored with the third party and in such case implementing a

log monitoring system to analyze the various logs related to a security breach and attacks could be a challenging task [116].

- In clouds, similar to dynamic resource provisioning, automatic resources provisioning is another challenging issue in which by predicting the future demand, resources are allocated and de-allocated from the cloud.
- Server consolidation, and energy efficiency is another big challenge in cloud computing to minimize the power consumption and operational cost of data centers. A key challenge in this is to achieve a good trade-off between energy savings and application performance [117].
- For some specific scenarios, cloud interoperability issue is an emerging challenge where it becomes difficult to integrate existing legacy systems with proprietary cloud APIs to get various cloud services [118].

Integration of cloud computing and IoT provides enormous benefits to the users and organizations in terms of more bandwidth and resources. Due to this reason, it has been widely adopted by the industry but still it has number of issues which need to be addressed and several more challenges are emerging related to applications security and privacy after integration of these technologies.

2.8 Summary

The advancement in computing field has gone through from traditional computing to interactive cloud and IoT-based predictive computing to make our lives easier and comfortable. The predictive computing makes the utilization of wireless sensor network for connectivity of various smart objects with Internet and continuous collection of data from these objects to make predictions related to health, navigation, agriculture, sales, etc. Predictive computing makes effective use of machine learning and data mining approaches to process collected data and produce the results in real time for consideration. However, these technological advancements that predictive computing brings are associated with security risks and privacy issues that need to be addressed thoroughly for effective implementation of predictive systems. Ignoring these risks and issues will adversely affect the system's integrity. In this chapter, we have presented a study on various predictive computing techniques and frameworks that can be applied in a variety of fields including vehicle navigation, sustainable computing, e-health, smart home and e-commerce, etc. As we know, integration of IoT and cloud computing sums up the total possible threats related to user and system in case of predictive computing. We have also presented the various information security techniques and frameworks related to cloud computing and IoT. These security techniques represent various security attacks like hidden attacks, eavesdropping, spoofing, etc., and security violations in terms of confidentiality, integrity, availability, trustworthiness, etc., of data. We have also outlined various challenges related to security and privacy issues of cloud computing and IoT. We hope that in future, researchers and developers will consider these security and privacy issues and will provide appropriate solutions to these issues at an early stage of system development. Predictive computing is the future that will change the way of application development scenario by changing the existing framework into the predictive framework and will provide short-term or long-term prediction results to enhance the day-to-day life of a user.

References

1. Hendrik H, Perdana DHF (2014) Trip guidance: a linked data based mobile tourists guide. Adv Sci Lett 20(1):75–79
2. Irudeen R, Samaraweera S (2013) Big data solution for Sri Lankan development: a case study from travel and tourism. In: Proceedings of international conference on advances in ICT for emerging regions (ICTer 2). IEEE, Colombo, pp 207–216
3. Dai L (2005) Fast shortest path algorithm for road network and implementation. http://people.scs.carleton.ca/~maheshwa/Honor-Project/Fall05-ShortestPaths.pdf. Accessed 15 Oct 2016
4. Kang WM, Moon SY, Park JH (2017) An enhanced security framework for home appliances in smart home. Hum Centric Comput Inf Sci 7(1):1–12
5. Malekian R, Kavishe AF, Maharaj BTJ, Gupta PK, Singh G, Waschefort H (2016) Smart vehicle navigation system using Hidden Markov model and RFID sensors. Wireless Pers Commun 90(4):1717–1742
6. Pattanaik V, Mayank S, Gupta, PK, Singh SK (2016) Smart real-time traffic congestion estimation and clustering technique for urban vehicular roads. In: Proceedings of IEEE region 10 conference (TENCON), Singapore, IEEE, pp 3420–3423
7. Gupta PK, Maharaj BTJ, Malekian R (2016) A novel and secure IoT based cloud centric architecture to perform predictive analysis of users activities in sustainable health centers. J Multimed Tools Appl. doi:10.1007/s11042-016-4050-6
8. Columbus L (2016) Roundup of internet of things forecasts and market estimates. https://www.forbes.com/sites/louiscolumbus/2016/11/27/roundup-of-internet-of-things-forecasts-and-market-estimates-2016/#17b3fabf292d. Accessed 10 Mar 2017
9. Kalechofsky H (2016) A simple framework for building predictive models. A Little Data Science Business Guide, pp 1–18
10. Zhu YH, Xu J, Li E, Xu L (2014) Energy-efficient reliable data gathering scheme based on enhanced reed-solomon code for wireless sensor networks. In: Proceedings of international conference on smart computing workshops (SMARTCOMP workshops), Hong Kong, IEEE, pp 275–280
11. Khan A, Imon SKA, Das SK (2014) Ensuring energy efficient coverage for participatory sensing in urban streets. In: Proceedings of international conference on smart computing workshops (SMARTCOMP workshops), Hong Kong, IEEE, pp 167–174
12. Abdullah S, Yang K (2014) An energy efficient message scheduling algorithm considering node failure in IoT environment. Wireless Pers Commun 79(3):1815–1835
13. Brienza S, Bindi F, Anastasi G (2014) e-net-manager: a power management system for networked PCs based on soft sensors. In: Proceedings of international conference on smart computing workshops (SMARTCOMP Workshops), Hong Kong, IEEE, pp 104–111
14. Gupta PK, Singh G (2015) A novel human computer interaction aware algorithm to minimize energy consumption. Wireless Pers Commun 81(2):661–683
15. Gupta PK, Singh G (2012) User centric framework of power schemes for minimizing energy consumption by computer systems. In: Proceedings of international conference on radar, communication and computing (ICRCC), India, IEEE, pp 48–53

16. Gupta PK, Singh G (2012) Energy-sustainable framework and performance analysis of power scheme for operating systems: a tool. Int J Intell Syst Appl 5(1):1–15
17. Gupta PK, Singh G (2011) A framework of creating intelligent power profiles in operating systems to minimize power consumption and greenhouse effect caused by computer systems. J Green Eng 1(2):145–163
18. Barth M, Karbassi A (2003) Vehicle route prediction and time of arrival estimation techniques for improved transportation system management. In: Proceedings of intelligent vehicles symposium, IEEE, pp 511–516
19. Froehlich J, Krumm J (2008) Route prediction from trip observations. SAE technical paper 2008-01-0201. doi:10.4271/2008-01-0201
20. Kansal A, Goraczko M, Zhao F (2007) Building a sensor network of mobile phones. In: Proceedings of 6th international conference on information processing in sensor networks (IPSN'07), ACM, pp 547–548
21. Suo H, Wan J, Li D, Zou C (2012) Energy management framework designed for autonomous electric vehicle with sensor networks navigation. In: Proceedings of 12th international conference on computer and information technology (CIT), Chengdu, Sichuan, China, IEEE, pp 914–920
22. Li Q, Chen L, Li M, Shaw SL, Nuchter A (2014) A sensor-fusion drivable-region and lane-detection system for autonomous vehicle navigation in challenging road scenarios. IEEE Trans Veh Technol 63(2):540–555
23. Chen M, Gonzalez S, Zhang Q, Leung VC (2010) Code-centric RFID system based on software agent intelligence. IEEE Intell Syst 25(2):12–19
24. Canino-Rodríguez JM, García-Herrero J, Besada-Portas J, Ravelo-García AG, Travieso-González C, Alonso-Hernández JB (2015) Human computer interactions in next-generation of aircraft smart navigation management systems: task analysis and architecture under an agent-oriented methodological approach. Sensors 15(3):5228–5250
25. Cao H, Wu W, Chen Y (2014) A navigation route based minimum dominating set algorithm in VANETs. In: Proceedings of international conference on smart computing workshops (SMARTCOMP workshops), Hong Kong, IEEE, pp 71–76
26. Davidson P (2013) Algorithms for autonomous personal navigation systems. https://tutcris.tut.fi/portal/files/2307019/davidson.pdf. Accessed 10 Apr 2017
27. Su JM, Chang CH, Yang TP, Chuang CF, Su SY (2014) Development of shortest path computing mechanism with consideration of commercial vehicles characteristics. In: Proceedings of international conference on smart computing workshops (SMARTCOMP workshops), Hong Kong, IEEE, pp 29–34
28. Mitton N, Rivano H (2014) On the use of city bikes to make the city even smarter. In: Proceedings of international conference on smart computing workshops (SMARTCOMP workshops), Hong Kong, IEEE, pp 3–8
29. Kranz M, Holleis P, Schmidt A (2010) Embedded interaction: interacting with the internet of things. IEEE Internet Comput 14(2):46–53
30. Huang W, Su X (2015) Design of a fault detection and isolation system for intelligent vehicle navigation system. Int J Navig Observ 2015(279086):19. doi:10.1155/2015/279086
31. Wang CC, Lien SF, Hsieh YC (2014) Integration of disaster detection and warning system for a smart vehicle. Adv Mech Eng 6:1–7
32. De Silva MWHM, Konara KMSM, Karunarathne IRAI, Lal KKUP, Wijesundara M (2014) An information system for vehicle navigation in congested road networks. SLIIT Res 113–116
33. Kim JH, Kim SC (2013) Design of architectural smart vehicle middleware. Information 16 (4):2443–2455
34. Wang C, Peng G (2015) Application of internet of things in development of e-navigation architecture. In: Proceedings of international symposium on computers and informatics (ISCI 2015). Atlantis Press, Beijing, China, pp 579–586
35. Wan J et al (2014) IoT sensing framework with inter-cloud computing capability in vehicular networking. Electron Commer Res 14(3):389–416

36. Huang YM, Chao HC, Park JH, Lai CF (2010) Adaptive body posture analysis for elderly-falling detection with multisensors. IEEE Intell Syst 25(2):20–30
37. Jeong YS, Song EH, Chae GB, Hong M, Park DS (2010) Large-scale middleware for ubiquitous sensor networks. IEEE Intell Syst 25(2):48–59
38. Taylor GA, Wallom DC, Grenard S, Huete AY, Axon CJ (2011) Recent developments towards novel high performance computing and communications solutions for smart distribution network operation. In: Proceedings of 2nd IEEE PES international conference and exhibition on innovative smart grid technologies (ISGT Europe), Manchester, UK, IEEE, pp 1–8
39. Gaoan G, Zhenmin Z (2014) Heart rate measurement via smart phone acceleration sensor. In: Proceedings of international conference on smart computing workshops (SMARTCOMP workshops), Hong Kong, IEEE, pp 295–300
40. Kortuem G, Kawsar F, Sundramoorthy V, Fitton D (2010) Smart objects as building blocks for the internet of things. IEEE Internet Comput 14(1):44–51
41. Ng K, Ghoting A, Steinhubl SR, Stewart WF, Malin B, Sun J (2014) PARAMO: a PARAllel predictive MOdeling platform for healthcare analytic research using electronic health records. J Biomed Inform 48:160–170
42. Brooks P, El-Gayar O, Sarnikar S (2015) A framework for developing a domain specific business intelligence maturity model: application to healthcare. Int J Inf Manage 35(3): 337–345
43. Lu R, Lin X, Shen X (2013) SPOC: a secure and privacy-preserving opportunistic computing framework for mobile-healthcare emergency. IEEE Trans Parallel Distrib Syst 24 (3):614–624
44. Zhang Y, Sun L, Song H, Cao X (2014) Ubiquitous WSN for healthcare: recent advances and future prospects. IEEE Internet Things J 1(4):311–318
45. Xu B, Da Xu L, Cai H, Xie C, Hu J, Bu F (2014) Ubiquitous data accessing method in IoT-based information system for emergency medical services. IEEE Trans Ind Inform 10 (2):1578–1586
46. Hu JX, Chen CL, Fan CL, Wang KH (2017) An intelligent and secure health monitoring scheme using IoT sensor based on cloud computing. J Sens 3734764. doi:10.1155/2017/3734764
47. Zhao Y, Healy BC, Rotstein D, Guttmann CR, Bakshi R, Weiner HL, Brodley CE, Chitnis T (2017) Exploration of machine learning techniques in predicting multiple sclerosis disease course. PLoS One 12(4):1–13
48. Abreu PH, Santos MS, Abreu MH, Andrade B, Silva DC (2016) Predicting breast cancer recurrence using machine learning techniques: a systematic review. ACM Comput Surv (CSUR) 49(3):1–40
49. Rana S, Gupta S, Phung D, Venkatesh S (2015) A predictive framework for modeling healthcare data with evolving clinical interventions. Stat Anal Data Mining ASA Data Sci J 8 (3):162–182
50. Sakr S, Elgammal A (2016) Towards a comprehensive data analytics framework for smart healthcare services. Big Data Res 4:44–58
51. Ifrim C, Pintilie AM, Apostol E, Dobre C, Pop F (2017) The art of advanced healthcare applications in big data and IoT systems. Advances in mobile cloud computing and big data in the 5G Era, pp 133–149
52. Mulvenna M, Nugent CD, Gu X, Shapcott M, Wallace J, Martin S (2006) Using context prediction for self-management in ubiquitous computing environments. In: Proceedings of consumer communications and networking conference, Nevada, USA, IEEE, pp 1–5
53. Apthorpe N, Reisman D, Feamster N (2017) A smart home is no castle: privacy vulnerabilities of encrypted IoT traffic. arXiv preprint arXiv:1705.06805, pp 1–6
54. Raj SV (2012) Implementation of pervasive computing based high-secure smart home system. In: Proceedings of international conference on computational intelligence and computing research (ICCIC), Coimbatore, India, IEEE, pp 1–8

55. Aquino-Santos R, Gonzalez-Potes A, Edwards-Block A, Garcia-Ruiz MA (2012) Ubiquitous computing and ambient intelligence for smart homes applications. In: Proceedings of world automation congress (WAC), Puerto Vallarta, Mexico, IEEE, pp 1–6
56. Hong Z, Li P, Jingxiao W (2013) Context-aware scheduling algorithm in smart home system. China Commun 10(11):155–164
57. Ning Y, Zhong-qin W, Malekian R, Ru-chuan W, Abdullah AH (2013) Design of accurate vehicle location system using RFID. Electron Elect Eng 40(8):105–110
58. Baum LE, Petrie T, Soules G, Weiss N (1970) A maximization technique occurring in the statistical analysis of probabilistic functions of Markov chains. Ann Math Stat 41:164–171
59. Simmons R, Browning B, Zhang Y, Sadekar V (2006) Learning to predict driver route and destination intent. In: Proceedings of conference on intelligent transportation systems (ITSX 06), IEEE, pp 127–132
60. Herbert SL, Chen M, Han S, Bansal S, Fisac JF, Tomlin CJ (2017) FaSTrack: a modular framework for fast and guaranteed safe motion planning. arXiv preprint arXiv:1703.07373, pp 1–8
61. Jabbarpour MR, Zarrabi H, Khokhar RH, Shamshirband S, Choo KKR (2017) Applications of computational intelligence in vehicle traffic congestion problem: a survey. Soft Comput 1–22
62. Yang JY, Chou LD, Tseng LM, Chen YM (2017) Autonomic navigation system based on predicted traffic and VANETs. Wireless Pers Commun 92(2):515–546
63. Cebecauer M, Jenelius E, Burghout W (2017) Integrated framework for real-time urban network travel time prediction on sparse probe data. https://people.kth.se/~jenelius/CJB_2017.pdf. Accessed 10 Mar 2017
64. Zhuang Y, Fong S, Yuan M, Sung Y, Cho K, Wong RK (2017) Predicting the next turn at road junction from big traffic data. J Supercomput 1–21. doi:10.1007/s11227-017-2013-y
65. Zhang M (2014) Path planning for autonomous vehicles. http://lib.dr.iastate.edu/cgi/viewcontent.cgi?article=5265&context=etd. Accessed 18 Feb 2017
66. Weiskircher T, Wang Q, Ayalew B (2017) Predictive guidance and control framework for (semi-) autonomous vehicles in public traffic. IEEE Trans Control Syst Technol 1–13
67. Lounis A (2015) Toward fully autonomous vehicle navigation using hybrid multi-controller architectures. Graduate theses. http://lounisadouane.online.fr/__Publications/LounisADOUANE_ManuscritHDR.pdf
68. Qiu J (2014) A predictive model for customer purchase behavior in e-commerce context. In: Proceedings of Pacific Asia conference on information systems (PACIS), pp 1–13
69. Gupta R, Pathak C (2014) A machine learning framework for predicting purchase by online customers based on dynamic pricing. Procedia Comput Sci 36:599–605 (Philadelphia, PA)
70. Ahmadi K (2011) Predicting e-customer behavior in B2C relationships for CLV model. Int J Bus Res Manage 2(3):128–138
71. Lo C, Frankowski D, Leskovec J (2016) Understanding behaviors that lead to purchasing: a case study of pinterest. In: Proceedings of 22nd ACM SIGKDD international conference on knowledge discovery and data mining, San Francisco, CA, USA, ACM, pp 531–540
72. Badea LM (2014) Predicting consumer behavior with artificial neural networks. Procedia Econ Finan 15:238–246
73. Naumzik C, Feuerriegel S, Neumann D (2017) Understanding consumer behavior in electronic commerce with image sentiment. In: Proceedings of 13th international conference on Wirtschaftsinformatik, St. Gallen, Switzerland, pp 1264–1266
74. Arockiam DL, Monikandan S (2014) A security service algorithm to ensure the confidentiality of data in cloud storage. Int J Eng Res Technol (IJERT) 3(12):1053–1058
75. Alsulami N, Alharbi E, Monowar MM (2015) A survey on approaches of data confidentiality and integrity models in cloud computing systems. J Emerg Trends Comput Inf Sci 6(3):188–197
76. Zhou L, Varadharajan V, Hitchens M (2013) Achieving secure role-based access control on encrypted data in cloud storage. IEEE Trans Inf Forensics Secur 8(12):1947–1960

77. Bokefode JD, Ubale Swapnaja A, Pingale Subhash V, Karande Kailash J, Apate Sulabha S (2015) Developing secure cloud storage system by applying AES and RSA cryptography algorithms with role based access control model. Int J Comput Appl 118(12):46–52
78. Gugnani G, Ghrera SP, Gupta PK, Malekian R, Maharaj BTJ (2016) Implementing DNA encryption technique in web services to embed confidentiality in cloud. In: Proceedings of second international conference on computer and communication technologies. Springer, Hyderabad, India, pp 407–415
79. Terec R, Vaida MF, Alboaie L, Chiorean L (2011) DNA security using symmetric and asymmetric cryptography. Int J New Comput Archit Appl (IJNCAA) 1(1):34–51
80. Grobauer B, Walloschek T, Stocker E (2011) Understanding cloud computing vulnerabilities. IEEE Secur Priv 9(2):50–57
81. Dinesha HA, Rao DH (2017) Evaluation of secure cloud transmission protocol. Int J Comput Netw Inf Secur 9(3):45–53
82. Yu Y, Au M Ho, Ateniese G, Huang X, Susilo W, Dai Y, Min G (2017) Identity-based remote data integrity checking with perfect data privacy preserving for cloud storage. IEEE Trans Inf Forensics Secur 12(4):767–778
83. Zhou J, Cao Z, Dong X, Vasilakos AV (2017) Security and privacy for cloud-based IoT: challenges. IEEE Commun Mag 55(1):26–33
84. Choi M, Lee C (2015) Information security management as a bridge in cloud systems from private to public organizations. Sustainability 7(9):12032–12051
85. Gaetani E, Aniello L, Baldoni R, Lombardi F, Margheri A, Sassone V (2017) Blockchain-based database to ensure data integrity in cloud computing environments. In: Proceedings of ITASEC, Venice, Italy, pp 146–155
86. Dabbagh M, Rayes A (2017) Internet of things security and privacy. Internet of things from hype to reality. Springer, pp 195–223
87. Nia AM, Jha NK (2017) A comprehensive study of security of internet-of-things. IEEE Trans Emerg Top Comput. doi:10.1109/TETC.2016.2606384
88. Gubbi J, Buyya R, Marusic S, Palaniswami M (2013) Internet of things (IoT): a vision, architectural elements, and future directions. Future Gener Comput Syst 29(7):1645–1660
89. Atzori L, Iera A, Morabito G (2010) The internet of things: a survey. Comput Netw 54 (15):2787–2805
90. CISCO (2014) The internet of things reference model. http://cdn.iotwf.com/resources/71/IoT_Reference_Model_White_Paper_June_4_2014.pdf. Accessed 11 Apr 2017
91. Singh S, Sharma PK, Moon SY, Park JH (2017) Advanced lightweight encryption algorithms for IoT devices: survey, challenges and solutions. J Ambient Intell Humaniz Comput. doi:10.1007/s12652-017-0494-4
92. Chang V, Kuo YH, Ramachandran M (2016) Cloud computing adoption framework: a security framework for business clouds. Future Gener Comp Syst 57:24–41
93. Mushtaq MO, Shahzad F, Tariq MO, Riaz M, Majeed B (2017) An efficient framework for information security in cloud computing using auditing algorithm shell (AAS). Int J Comput Sci Inf Secur (IJCSIS) 14(11):317–331
94. Youssef AE, Alageel M (2012) A framework for secure cloud computing. IJCSI Int J Comput Sci Issues 9(4):487–500
95. Matte V, Kumar LR (2013) A new framework for cloud computing security using secret sharing algorithm over single to multi-clouds. Int J Comput Trends Technol (IJCTT) 4 (8):2820–2824
96. Dorairaj SD, Kaliannan T (2015) An adaptive multilevel security framework for the data stored in cloud environment. Sci World J 1–11
97. Ondiege B, Clarke M, Mapp G (2017) Exploring a new security framework for remote patient monitoring devices. Computers 6(11):1–12
98. Jaganathan S, Veerappan D (2015) CIADS: a framework for secured storage of patients medical data in cloud. Int J WSEAS Trans Inf Sci Appl 12:22–35
99. Xiao Z, Xiao Y (2013) Security and privacy in cloud computing. IEEE Commun Surv Tutor 15(2):843–859

100. Ning H, Liu H, Yang LT, Cyberentity security in the internet of things. Computer 46(4): 46–53
101. Vermesan O, Friess P, Guillemin P, Gusmeroli S, Sundmaeker H, Bassi A, Jubert IS, Mazura M, Harrison M, Eisenhauer M, Doody P (2011) Internet of things strategic research roadmap. Internet Things-Global Technol Soc Trends 1:9–52
102. Roman R, Zhou J, Lopez J (2013) On the features and challenges of security and privacy in distributed internet of things. Comput Netw 57(10):2266–2279
103. Park KC, Shin DH (2017) Security assessment framework for IoT service. Telecommun Syst 64(1):193–209
104. Ge M, Hong JB, Guttmann W, Kim DS (2017) A framework for automating security analysis of the internet of things. J Netw Comput Appl 83:12–27
105. Ngu AH, Gutierrez M, Metsis V, Nepal S, Sheng QZ (2017) IoT middleware: a survey on issues and enabling technologies. IEEE Internet Things J 4(1):1–20
106. Ukil A, Bandyopadhyay S, Pal A (2015) Privacy for IoT: involuntary privacy enablement for smart energy systems. In: Proceedings of international conference on communications (ICC), IEEE, pp 536–541
107. TrustArc (2015) 35% of Americans now own at least one smart device other than a phone. https://www.trustarc.com/press/35-of-americans-now-own-at-least-one-smart-device-other-than-a-phone/. Accessed 8 Mar 2017
108. Marium S, Nazir Q, Ahmed A, Ahthasham S, Mirza AM (2012) Implementation of EAP with RSA for enhancing the security of cloud computing. Int J Basic Appl Sci 1(3):177–183
109. Wang J, Zhao Y, Jiang S, Le J (2009) Providing privacy preserving in cloud computing. In: Proceedings of international conference on test and measurement (ICTM 2009), vol 2. Hong Kong, China, IEEE, pp 213–216
110. Greveler U, Justus B, Loehr D (2011) A privacy preserving system for cloud computing. In: Proceedings of 11th international conference on computer and information technology, IEEE, pp. 648–653
111. Zhou M, Mu Y, Susilo W, Au MH, Yan J (2011) Privacy-preserved access control for cloud computing. In: Proceedings of 10th international conference on trust, security and privacy in computing and communications (TrustCom), Changsha, China, IEEE, pp 83–90
112. Rahaman SM, Farhatullah M (2012) PccP: a model for preserving cloud computing privacy. In: Proceedings of international conference on data science and engineering (ICDSE), Cochin, Kerala, India, IEEE, pp 166–170
113. Wang C, Chow SS, Wang Q, Ren K, Lou W (2013) Privacy-preserving public auditing for secure cloud storage. IEEE Trans Comput 62(2):362–375
114. Wang B, Li B, Li H (2012) Oruta: privacy-preserving public auditing for shared data in the cloud. In: Proceedings of 5th international conference on cloud computing (CLOUD), Honolulu, HI, USA, IEEE, pp 295–302
115. NIST (2014) Framework for improving critical infrastructure cybersecurity. https://www.ncjrs.gov/App/Publications/abstract.aspx?ID=267567. Accessed 14 Apr 2017
116. Kent K, Souppaya M (2006) Guide to computer security log management. US Department of Commerce, National Institute of Standard and Technology, Gaithersburg, MD, USA, p 16
117. Zhang Q, Cheng L, Boutaba R (2010) Cloud computing: state-of-the-art and research challenges. J Internet Serv Appl 1(1):7–18
118. Dillon T, Wu C, Chang E (2010) Cloud computing: issues and challenges. In: Proceedings of 24th international conference on advanced information networking and applications (AINA), Perth, Western Australia, IEEE, pp 27–33
119. Wang C, Zhang Y, Song WZ (2014) A new data aggregation technique in multi-sink wireless sensor networks. In: Proceedings of international conference on smart computing workshops (SMARTCOMP workshops), Hong Kong, IEEE, pp 99–104
120. Villari M, Celesti A, Fazio M, Puliafito A (2014) Alljoyn lambda: an architecture for the management of smart environments in IoT. In: Proceedings of international conference on smart computing workshops (SMARTCOMP workshops), Hong Kong, IEEE, pp 9–14

121. Zhu W, Cui X, Hu C, Ma C (2014) Complex data collection in large-scale RFID systems. In: proceedings of international conference on smart computing workshops (SMARTCOMP workshops), Hong Kong, IEEE, pp 25–32
122. Wang C, Peng Y, De D, Song WZ (2014) DCTP: data collecting based on trajectory prediction in smart environment. In: Proceedings of international conference on smart computing workshops (SMARTCOMP workshops), Hong Kong, IEEE, pp 93–98
123. Sharma K, Singh KR (2013) Seed block algorithm: a remote smart data back-up technique for cloud computing. In: Proceedings of international conference on communication systems and network technologies (CSNT), Gwalior, India, IEEE, pp 376–380
124. Flinsenberg ICM (2004) Route planning algorithms for car navigation. http://brainmaster.com/software/pubs/brain/Flinsenberg%20Route%20Planning.pdf. Accessed 17 Apr 2017
125. Parulekar M, Padte V, Shah T, Shroff K, Shetty R (2013) Automatic vehicle navigation using Dijkstra's algorithm. In: Proceedings of international conference on advances in technology and engineering (ICATE), Mumbai, India, IEEE, pp 1–5
126. Fu M, Li J, Deng Z (2004) A practical route planning algorithm for vehicle navigation system. In: Proceedings of fifth world congress on intelligent control and automation (WCICA), vol 6. Hangzhou, China, June, IEEE, pp 5326–5329
127. Eppstein D (1998) Finding the k shortest paths. SIAM J Comput 28(2):652–673
128. Lebedev A, Lee J, Rivera V, Mazzara M (2017) Link prediction using top-k shortest distances. arXiv preprint arXiv:1705.02936, pp 1–5
129. Shahzada A, Askar K (2011) Dynamic vehicle navigation: an A* algorithm based approach using traffic and road information. In: Proceedings of international conference on computer applications and industrial electronics (ICCAIE), Penang, Malaysia, IEEE, pp 514–518
130. Goldberg AV, Radzik T (1993) A heuristic improvement of the Bellman-Ford algorithm. Appl Math Lett 6(3):3–6
131. Salehinejad H, Nezamabadi-pour H, Saryazdi S, Farrahi-Moghaddam F (2008) Combined A*-ants algorithm: a new multi-parameter vehicle navigation scheme. In: Iranian conference on electrical engineering (ICEE 2008), Tehran, Iran, IEEE, pp 154–159
132. Rahaman MS, Mei Y, Hamilton M, Salim FD (2017) CAPRA: a contour-based accessible path routing algorithm. Inf Sci 385:157–173
133. Zhao L, Ochieng WY, Quddus MA, Noland RB (2003) An extended Kalman filter algorithm for integrating GPS and low cost dead reckoning system data for vehicle performance and emissions monitoring. J Navig 56(2):257–275
134. Hu C, Chen W, Chen Y, Liu D (2003) Adaptive Kalman filtering for vehicle navigation. J Glob Pos Syst 2(1):42–47
135. Jin B, Guo J, He D, Guo W (2017) Adaptive Kalman filtering based on optimal autoregressive predictive model. GPS Solut 21(2):307–317
136. Ko E, Kang J, Park J (2012) A middleware for smart object in ubiquitous computing environment. In: Proceedings of 8th international conference on computing technology and information management (ICCM), Seoul, Korea (South), IEEE, pp 400–403
137. Solanas A, Patsakis C, Conti M, Vlachos IS, Ramos V, Falcone F, Postolache O, Pérez-Martínez PA, Di Pietro R, Perrea DN, Martinez-Balleste A (2014) Smart health: a context-aware health paradigm within smart cities. IEEE Commun Mag 52(8):74–81
138. Bottazzi D, Montanari R, Toninelli A (2007) Context-aware middleware for anytime, anywhere social networks. IEEE Intell Syst 22(5):23–32
139. Soliman M, Abiodun T, Hamouda T, Zhou J, Lung CH (2013) Smart home: integrating internet of things with web services and cloud computing. In: Proceedings of 5th international conference on cloud computing technology and science (CloudCom), vol 2. Bristol, UK, IEEE, pp 317–320
140. Siebert J, Cao J, Lai Y, Guo P, Zhu W (2015) LASEC: a localized approach to service composition in pervasive computing environments. IEEE Trans Parallel Distrib Syst 26 (7):1948–1957
141. Salmani H, Tehranipoor MM (2016) Vulnerability analysis of a circuit layout to hardware Trojan insertion. IEEE Trans Inf Forensics Secur 11(6):1214–1225

142. Shila DM, Venugopal V (2014) Design, implementation and security analysis of hardware Trojan threats in FPGA. In: Proceedings of international conference on communications, Sydney, Australia, IEEE, pp 719–724
143. Wehbe T, Mooney VJ, Keezer DC, Parham NB (2015) A novel approach to detect hardware Trojan attacks on primary data inputs. In: Proceedings of WESS'15: workshop on embedded systems security, Amsterdam, Netherlands, ACM, pp 1–10
144. Becher A, Benenson Z, Dornseif M (2006) Tampering with motes: real-world physical attacks on wireless sensor networks. In: Proceedings of international conference on security in pervasive computing. Springer, New York, UK, pp 104–118
145. Anderson R, Kuhn M (1996) Tamper resistance-a cautionary note. In: Proceedings of second Usenix workshop on electronic commerce, vol 2. Oakland, California, pp 1–11
146. Zorzi M, Gluhak A, Lange S, Bassi A (2010) From today's intranet of things to a future internet of things: a wireless-and mobility-related view. IEEE Wirel Commun 17(6):44–51
147. El Beqqal M, Azizi M (2017) Classification of major security attacks against RFID systems. In: Proceedings of international conference on wireless technologies, embedded and intelligent systems (WITS), Fez, Morocco, IEEE, pp 1–6
148. Uwagbole SO, Buchanan WJ, Fan L (2017) Applied machine learning predictive analytics to SQL injection attack detection and prevention. In: Proceedings of 3rd IEEE/IFIP workshop on security for emerging distributed network technologies (DISSECT), Lisbon, Portugal, IEEE, pp 1–4
149. Hong K, Lillethun D, Ramachandran U, Ottenwälder B, Koldehofe B (2013) Mobile fog: a programming model for large-scale applications on the internet of things. In: Proceedings of 2nd SIGCOMM workshop on mobile cloud computing, Hong Kong, China, ACM, pp 15–20
150. Martin T, Hsiao M, Ha D, Krishnaswami J (2004) Denial-of-service attacks on battery-powered mobile computers. In: Proceedings of 2nd conference on pervasive computing and communications (PerCom), Orlando, Florida, IEEE, pp. 309–318
151. Agah A, Das SK (2007) Preventing DoS attacks in wireless sensor networks: a repeated game theory approach. IJ Network Secur 5(2):145–153
152. C′ardenas AA, Amin S, Lin ZS, Huang YL, Huang CY, Sastry S (2011) Attacks against process control systems: risk assessment, detection, and response. In: Proceedings of 6th symposium on information, computer and communications security, ACM, pp 355–366
153. Mukherjee A (2015) Physical-layer security in the internet of things: sensing and communication confidentiality under resource constraints. Proc IEEE 103(10):1747–1761
154. Revathi B, Geetha D (2012) A survey of cooperative black and gray hole attack in MANET. Int J Comput Sci Manage Res 1(2):205–208
155. Heer T, Garcia-Morchon O, Hummen R, Keoh SL, Kumar SS, Wehrle K (2011) Security challenges in the IP-based internet of things. Wirel Pers Commun 61(3):527–542
156. Wallgren L, Raza S, Voigt T (2013) Routing attacks and countermeasures in the RPL-based internet of things. Int J Distrib Sens Netw 9(8):1–11
157. Barreno M, Nelson B, Sears R, Joseph AD, Tygar JD (2006) Can machine learning be secure. In: Proceedings of symposium on information, computer and communications security, Taipei, Taiwan, ACM, pp 16–25
158. Huang L, Joseph AD, Nelson B, Rubinstein BI, Tygar JD (2011) Adversarial machine learning. In: Proceedings of 4th ACM workshop on security and artificial intelligence, Chicago, IL, USA, ACM, pp 43–58

Chapter 3
Predictive Computing: A Technical Insight

3.1 Introduction

Predictive analytics is used to derive information in real time from existing datasets for finding patterns and predicting outcomes of existing problems. Predictive analytics encompasses of the advanced analytics and statistical techniques which come from predictive modelling, data mining, machine learning and artificial intelligence. Also, predictive analytics analyzes the datasets to make predictions about future trends. Datasets may be generated from various heterogeneous sources based on the current and previous facts. The patterns could be derived from existing dataset on the basis of the previous historical and transactional data and also to identify various security risks and attacks in near future. These predictive models analyze the relationships among many factors and present the various set of conditions. This set of conditions helps to assign a weight which contributes to assessing the risk. On successful implementation of predictive models, organizations can easily interpret the big data to gain maximum benefit. These predictive models work in association with data mining and text analytics and create predictive intelligence by finding hidden patterns and relationships in the dataset that consist data in structured and unstructured manner. Unstructured data represents the data collected from different sources like call center notes, social media content, or from a different type of open text which is extracted from the main text. The data which can be used readily for analysis is structured data, e.g. sales, income, age, gender and marital status. The sentiment is also extracted along with the text to build the model.

Currently, Predictive Analytics is also an emerging area related to big data. It has been gaining more proliferations in recent years [1, 2]. Despite this current majority and popularity, the underlying data is at least 20–30 years old. Many researchers, engineers and scientists have applied the predictive models for interpretation of the big data and prediction [3]. The definition of big data concerns with massive volume, complex and growing data sets with multiple and independent sources

P.K. Gupta et al., *Predictive Computing and Information Security*,
DOI 10.1007/978-981-10-5107-4_3

[4, 5]. Accompanying with the above example, the era of big data has appeared [4–6]. Every day, nearly 2.5 quintillion bytes of data are generated from the different websites, autonomous sources, emerging ubiquitous devices. 93% of the data in the world today is created within the past few years [5–7]. In the direction of big data generation, social websites, online shopping websites such as Facebook, Yahoo, Amazon, tutor Flickr, etc., have got proliferation and received 1.8 million photos per day on an average during 2015–2016 [8]. This requires nearly 100 terabytes (TB) storage space every day. The billions of pictures on Flicker and other social sites are included as treasure tank to explore and retrieval of data items of the human society, social cultural events, public discussions and affairs, awareness of disasters, genetically encoding and decoding of gene expressions and so on, only if human society have the power to harness the massive amount of data [8–12].

These examples provide a good demonstration of the rise of big data applications. The data collection has grown massive and tremendously in the recent time. It is beyond the ability of commonly used traditional computing paradigm, statistical programme's applications and software tools to capture, manage and process this massive amount of big data within a defined tolerable elapsed time. The major challenges for big data applications are to explore the massive volumes of data and to extract discriminatory information or knowledge for future trends [8, 9, 12, 13]. In many cases, knowledge extraction is done in real time or near real time so it could be difficult to store all the observed data. As a result, the unprecedented data volumes require an effective data analysis and prediction platform to achieve fast response and real-time classification for such big data [10]. The recent research in the modern big data and data sciences provides new paradigms for computation and handling for the massive amount of data and learning approaches aimed at effectually searching through the enormous volume of data [10–12, 14].

Predictive analytics in big data is recently one of the active research areas because of its wide proliferation, application and uses in different research fields such as online photo-sharing like Flickr, social search, online video-sharing sites similar to YouTube, monitoring of ubiquitous computing, smart device, IoT-based intelligent systems and communication over internet services [10, 11, 13, 15, 16]. However, due to the ever-growing size of the web services, simple text-based communication or classical techniques for visual feature-based approaches for retrieving meaningful information from huge data may not be sufficient for optimal search results [13, 14, 17]. Data science also prompted multidisciplinary researchers to use various mathematical models, tools, technical innovations like big data analytics, predictive analytics, data mining concepts, coherent set of machine learning algorithms, deep learning algorithms and pattern recognition techniques to extract knowledge of the data information and help in decision making. With the promises of predictive analytics and computing in big data, the use of machine learning and pattern recognition algorithms predicting future are no longer challenging [5, 6, 18], especially for e-transportation, health monitoring, medical sciences, smart sensors and prediction of incurable diseases as predictive analytics play a vital role in the analysis and prediction of massive amount of data sufficiently. Predictive analytics and computing paradigms can be applied to data science [4], for analysis of customer behaviour to predict market value, to enhance financial services, to promote telecommunications, analysis of retail, object mobility, healthcare [19] and other fields [11, 20].

3.2 Design Architecture of Predictive Computing

3.2.1 Predictive Analytics Process

Predictive analytics process is the complete study from data collection to prediction. It is a cyclic process as shown in Fig. 3.1. This process starts from problem definition where problem statement is defined then data is collected based on problem definition. Further preprocessing techniques like data cleaning, data transformation are applied to make data appropriate for predictive modelling. Finally, deployment of prediction system takes place.

- *Problem definition*: In this segment, following project outcomes, scope, deliverables and business objectives are identified. The dataset to be used is identified and properly defined.
- *Data collection*: Dataset is prepared by collecting data from various sources that may be homogeneous or heterogeneous, that can provide a complete view of prediction pattern or trends.
- *Data preprocessing*: is the way of finding useful information by cleaning, transforming, inspecting and modelling data.

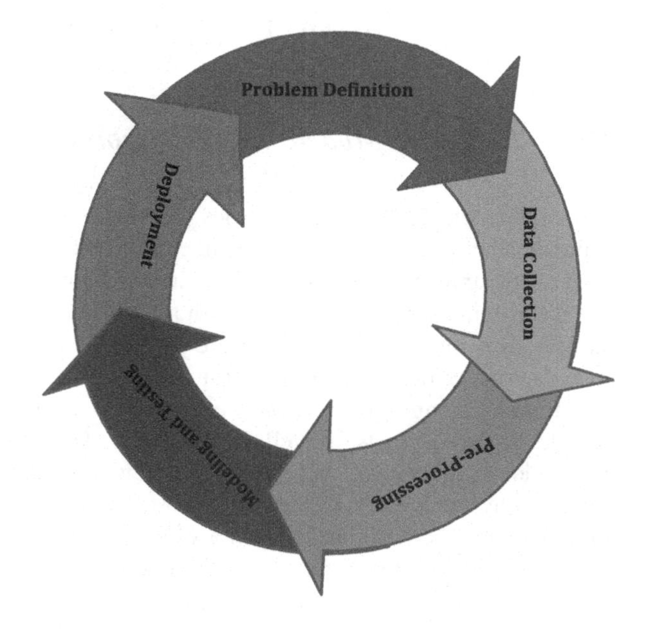

Fig. 3.1 Predictive analytics

- *Modelling and testing*: is the process of creating a predictive model that provides suitable probability of an outcome.
- *Deployment*: Predictive deployment model gives the option to deploy the result into the everyday decision making process to automate the process of results and results based on applied modelling techniques.

3.3 Predictive Model

The term predictive analytics is defined as a technique to predict the various kinds of the patterns or behaviours patterns of the massive amount of data. It is applied as predictive modelling-based systems. The basic working of predictive analytics paradigm is based on a coherent set of concepts of statistical techniques, pattern recognition and data mining algorithms [21]. It extracts a discriminatory set of information from the massive amount of data and performs the prediction and analysis of unknown data with the trained model. It predicts different types of trends and behaviour patterns of unknown data [22].

Predictive analytics is associated with the use of predictive modelling, forecasting and scoring data with predictive models. However, this term can be used to refer analytical disciplines, such as descriptive modelling and decision modelling or optimization. In statistics, a model is defined as a representative model of a relationship between variables (independent or dependent variable) in the data. The representational model illustrates how one variable or more variables are related to other variables. It also shows the changes in the behaviour patterns in the given data [23, 24]. The basic working principle of the model is a process in which an illustrative abstraction is developed from the set of observed data.

3.3.1 Predictive Model for e-Transportation

Urbanization in developing countries like India and China has led to massive population burst in their metro cities like New Delhi and Beijing. Mobility in these cities or the transport network within cities tremendously impacts both the city's and the nation's socio economic growth [25]. Due to increase in vehicles on roads, traffic congestion has become a severe problem in urban areas. Traffic congestion constrains the growth of Gross Domestic Product (GDP) of any developing country. Due to traffic congestion various problems arise such as increase in air pollution, vehicle operating costs, travelling time, etc. According to a report [26], Rs. 600 billion are lost every year due to traffic jams or delays on high volume roads and highways which also includes fuel wastage. Hence, it is obligatory that systems and algorithms be designed in such a way that people are able to avoid the traffic congestions in real time. Proposed e-Transportation system [27] is based on Global

Positioning System (GPS) enabled mobile phones which are abundantly available making the system particularly suitable for developing countries. Proposed real-time congestion avoidance method highlights prediction of the shortest route based on *k-means* Clustering of traffic data points. The proposed methodology can predict which road segments are congested or cleared through real-time GPS data. The system informs the driver about real-time traffic conditions and adjusts the route so as to avoid congestions and reduce travelling time drastically.

– ***e-Transportation Methodology***

The major objective of e-Transportation system is to identify the shortest path from the driver's current location (source) to the destination avoiding any traffic congestions in the best possible fashion. To accomplish this objective, the proposed e-Transportation system identifies road networks of a particular area using Google Maps. The system then converts the road network into a weighted graph where each intersection is denoted by a node and each edge represents a road. Based on the weighted graph, a neighbourhood metric is generated with all intersecting nodes; further applying Dijkstra's algorithm to discover the shortest path (with minimum traffic congestion) from source to destination. The proposed real-time traffic congestion avoidance system as shown in Fig. 3.2 is primarily composed of five basic steps:

(a) Fetch road map of the driver's vicinity. Assume the driver's current location is the source, and ask the user to feed in his/her destination.
(b) Identify the shortest path from source to destination in ideal condition, i.e. without traffic.
(c) Collect real-time GPS data of vehicles (through GPS-based mobile phone app) and plot the real-time traffic data onto the road map.
(d) Use the real-time traffic data attained in the previous step to create traffic clusters using *k-means Clustering* algorithm.
(e) Formulate the density of traffic clusters and identify the alternate shortest routes to avoid congestion. Reiterate the process in real time until the driver reaches the destination.

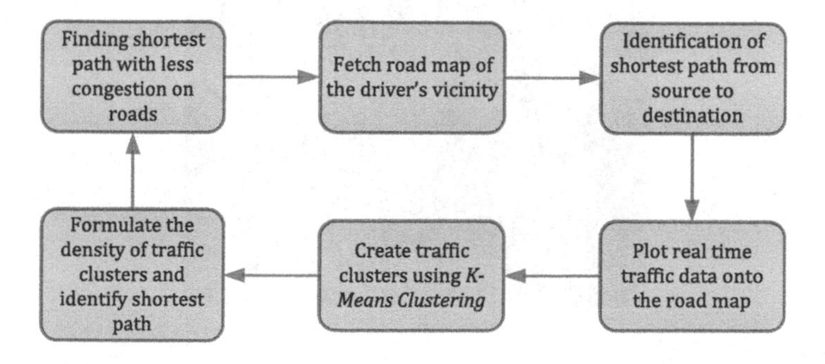

Fig. 3.2 Predictive model for finding shortest route with less traffic congestion on roads

– ***Road map to graph conversion***

The e-Transportation system fetches the road map of driver's local area from sources like Google Maps, Open Street Map, etc. The labelled map is processed using image processing techniques to fetch road map data only as shown in Fig. 3.3. The system further converts the road map image ($n \times n$ matrix) into neighbourhood matrix ($n^2 \times n^2$ matrix) where each element represents the relationship with its neighbouring pixels. This matrix represents whether a directly connected road is available or not. Weights of all directly connected nodes are initialized to 1 in the initial road network because different vehicles may pass through these road segments. The road network can be represented as a weighted graph $G = (V, E)$, where each intersection is denoted by a node V and each edge

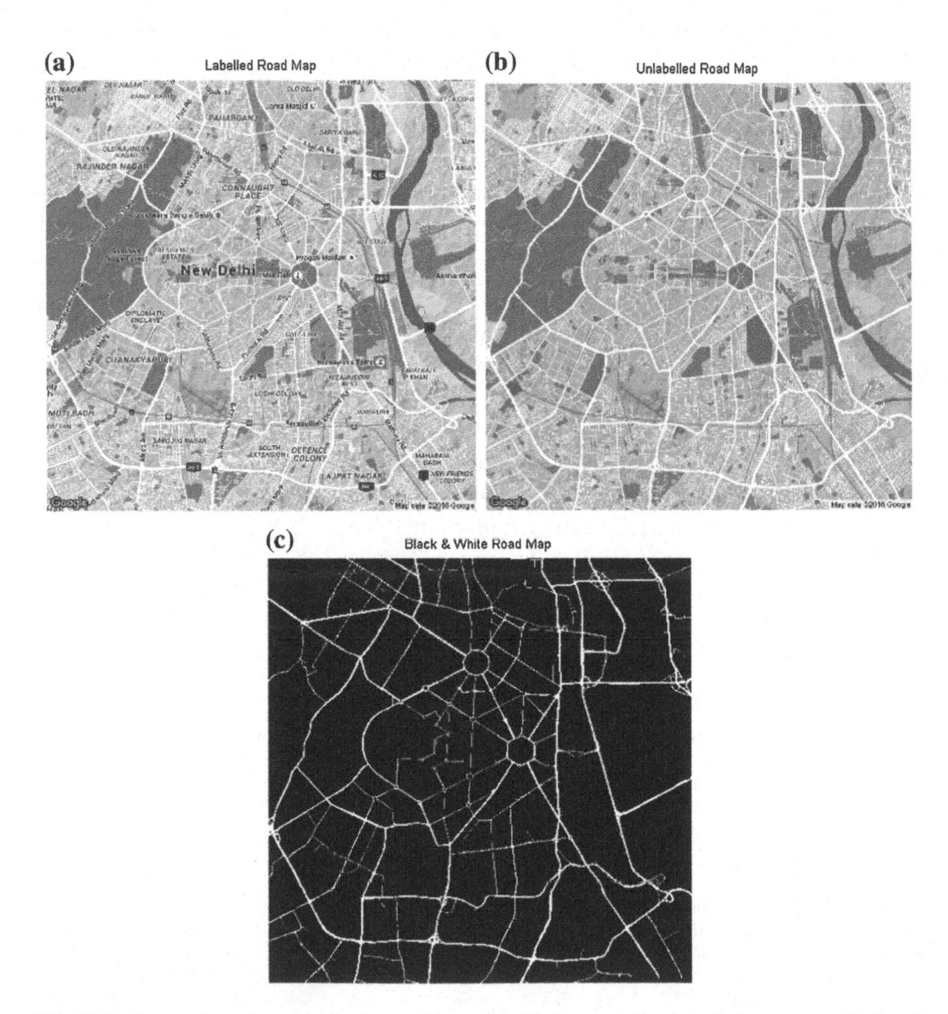

Fig. 3.3 Conversion of google road map from **a** labelled map to **b** unlabelled map to **c** *black* and *white* 2D matrix

E represents a road in the graph. When a vehicle is driven from a source S to destination D, an ordered set of roads in the route is defined as $A_i = (a_1; a_2; a_3; \ldots; a_n)$ where A_i represents the ith road and n is a total number of roads. According to graph theory, every route can be represented as $A_i = (e_1 \rightarrow v_1, v_2; e_2 \rightarrow v_2, v_3; \ldots; e_n \rightarrow v_n, v_{n+1})$.

- ***Traffic clustering and congestion estimation***

To get the current traffic condition of driver's local area, the system uses real-time GPS tracking of vehicle users using android application. The GPS location data transmitted by the vehicles gets collected by local clouds. The real-time data transmitted by an App includes vehicle location (longitude and latitude), speed, drive time, direction, etc. An institution-based proof of concept Android App was developed to do the same, although there are few alternative applications or services available on the internet which can be exploited to accomplish the above-mentioned tasks.

The GPS data with vehicles locations received from the application is fed into a two-dimensional problem space of size $n \times n$. The vehicle data is plotted on to driver's local road map. k-means clustering is then implemented so as to create traffic clusters with minimum Euclidean distance. Higher number of vehicle points within a cluster would then denote higher density or more congestion. Based on similarity or dissimilarity metric, clustering in N-dimensional Euclidean space is the process of partitioning a given set of N points into m clusters. Let, a set S is representing n points $(n_1, n_2, n_3, \ldots, n_n)$ and m clusters be representing $(a_1, a_2, a_3, \ldots, a_m)$ then (Eq. 3.1),

$$\begin{aligned} & A_i \neq \Phi \quad \text{for} \quad (i = 1, 2, 3, \ldots, m) \\ & A_i \cap A_j \quad \text{for} \quad (i = 1, 2, 3, \ldots, m); \\ & \text{Here,} \quad (j = 1, 2, 3, \ldots, m) \text{ and } i \neq j \\ & \bigcup_{i,j=1}^{m} A_i, A_j = S \end{aligned} \tag{3.1}$$

The road network is represented as neighbourhood matrix where each intersection is represented by a node and each edge represents a road segment. The initial weights are assumed to be 1, if two vertices' are directly connected with each other which represents the estimated travel time. When the GPS data is plotted on road segments to show the availability of vehicles on a particular road, the weights are updated periodically according to every vehicle available on each road segment. These updated weights are used to estimate the travel time or, to calculate the shortest path from current location to the desired destination with congestion avoidance. To estimate the travel time Greenshield's model is used. The model considers that there is a linear relationship between vehicle per metre traffic density D_i on road i and estimated road speed S_i. One can formulate this situation according to following formula (Eq. 3.2):

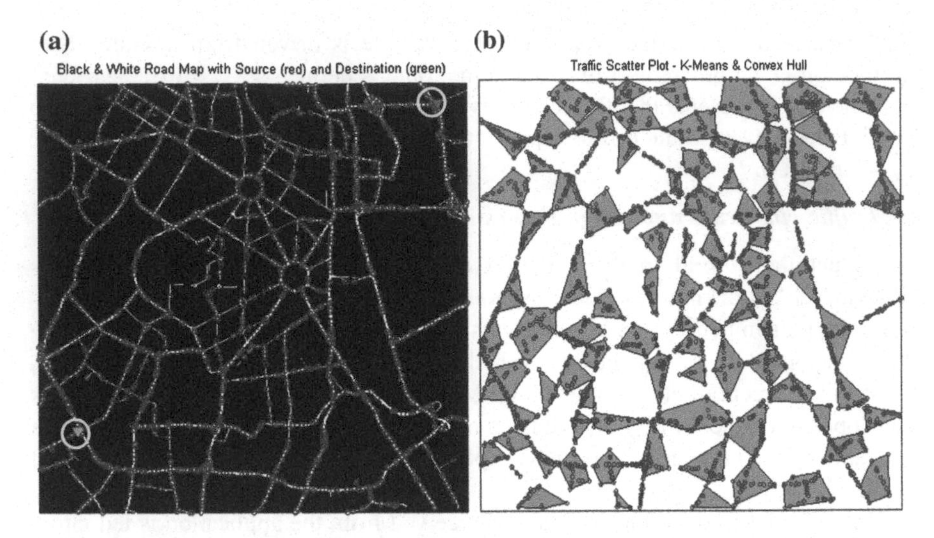

Fig. 3.4 *Red*, *green* and *blue points* on **a** *black* and *white* road map represent driver's location, destination and other vehicles, respectively **b** visualization of convex hulls for traffic clusters

$$S_i = S_j\left(1 - \frac{D_i}{D_j}\right) \tag{3.2}$$

where D_j is traffic density in jam and, S_i represents the speed flow on ith road segment (Eq. 3.3)

$$A_i = \frac{P_i}{S_i} \tag{3.3}$$

where A_i is the estimated travel time and P_i is the path length of the same road segment. Road density is calculated with Convex Hull algorithm on each identified *k-means* cluster. Dataset of vehicles available on road is provided to the algorithm to create convex hulls. Convex hulls represent the traffic congestion on particular road segment as shown in Fig. 3.4.

Now with the help of Dijkstra's algorithm, the system finds an alternate shortest path P'_i to avoid the congested roads and save travelling time. During the alternate path finding, road segments with higher weight in neighbourhood matrix are ignored to avoid the congestion and estimate the new travel time with the alternate path (Eq. 3.4).

$$A_i = \frac{P'_i}{S_i} \tag{3.4}$$

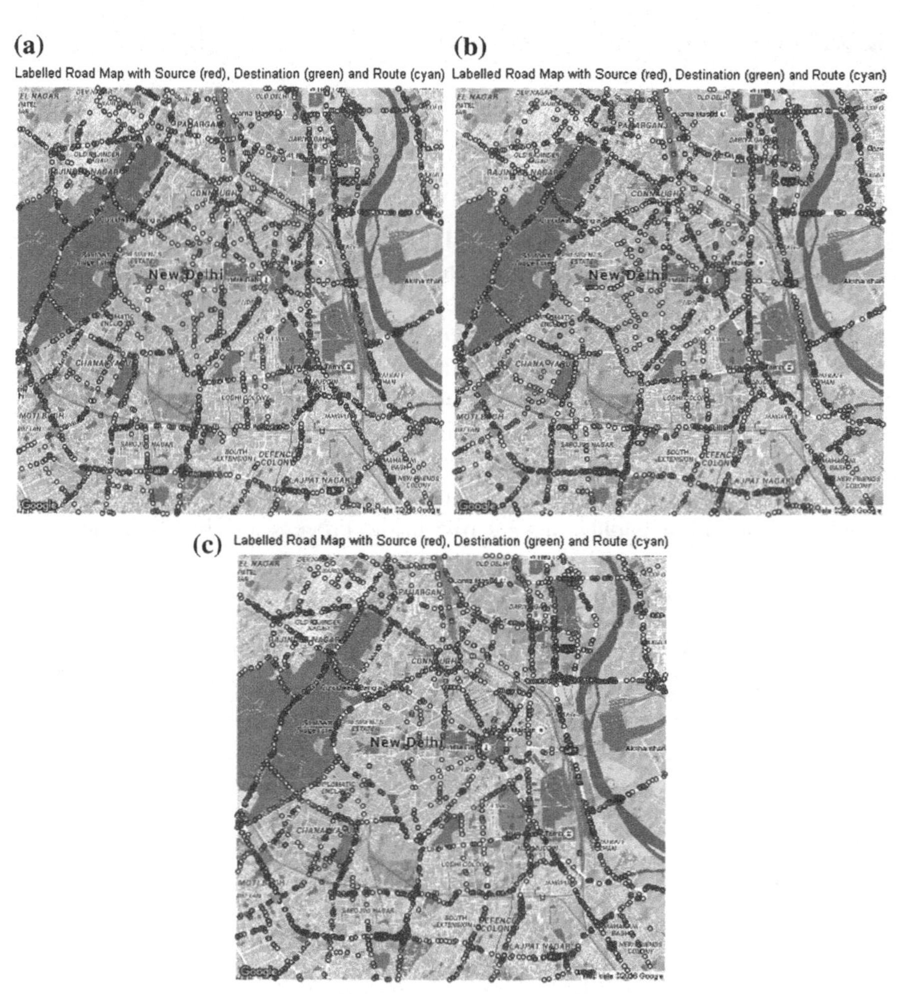

Fig. 3.5 *Red*, *green*, *blue* and *cyan points* on road map represent driver's location, destination, other vehicles and predicted path, respectively. The path changes dynamically with driver's location, starting from **a** position p_1 to **b** position p_r, and **c** position p_s

Figure 3.5 shows predicted routes from different positions p_1, p_r and p_s. As it can be seen in the figure, the routes predicted by the system are dynamic and change with traffic density on various roads from source to destination.

3.3.2 Predictive Model for Banking

In banking sectors, the predictive computing-based model predicts the customers' needs to be based on their money requirements. The predictive models extract

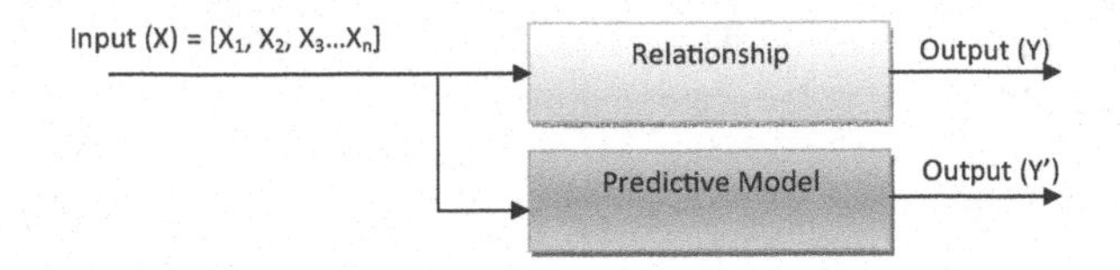

Fig. 3.6 Working of a predictive model

relevant information from customer data. It calculates the credit score, income level and requested loan amount to ascertain the basic interest rate of the loan [28].

For this work, it is required to have observational data (customer data) with the calculated credit score, loan amount, income level and interest rate of respective banks. Figure 3.6 illustrates working of a predictive model. The predictive computing model takes input from the set of observational data based on generalization techniques. Based on provided data, representative model can predict the value of the bank interest rate, based on all the input values $[X] = [X_1, X_2, X_3, \ldots, X_n]$. Here, X_1 = credit score, X_2 = income level and X_3 = loan amount, etc. In predictive computing, data mining plays a vital role in the process and building the representative model.

The predictive model works in following ways: (1) It predicts the output based on provided inputs (calculation of interest rate) and (2) It can be applied to understand the inferences and relationship between the variables (output variable and all the input variables) as shown in Fig. 3.6. Therefore, the building of predictive models based on machine learning and statistical techniques can be applied in both predictive analysis and critical applications such as e-health monitoring, security issues, resource utilization of clouds, public awareness, e-governance and diagnosis of dangerous diseases in medical fields [29, 30], etc.

3.4 Algorithms for Predictive Computing

To convert predictive model into a working machine, predictive algorithms play an important role. These algorithms are implemented on stored big data for further extraction of knowledge. In this section, we have discussed some major predictive computing algorithms, their advantages and disadvantages.

3.4.1 Local Learning and Model Fusion for Multiple Information Sources

Big data applications are mostly highlighted with independent sources, decentralized controls and systems and aggregating distributed data sources to a centralized site for mining is systematically prohibitive due to its high transmission cost and privacy concerns like issues [23, 31–33]. However, analysis and data mining-related activities can be carried out at each distributed sites around the globe, but this activity leads to biased view of the data collected at each location [34–36]. Under such a condition, a big data mining system has to facilitate a knowledge exchange and fusion mechanism to guarantee that all shared sites or sources can work collectively and mutually to achieve a global optimization objective [23, 34, 37–39]. More specifically, the global mining can be emphasized with a two-step correlation. It is known as local mining-based correlation and global mining-based correlation process, at data, model and at knowledge levels. Further, at the data level, each local systems can determine the data statistics and relations based on the local information of given data sources [40]. At the system model or pattern level, each control site can give the local mining set of activities, on the different localized data, to discover local discriminatory patterns [36, 40–42]. At the information level, model correlation analysis investigates the relevance of models generated from various data sources to decide how important the data sources are correlated with each other, and how to form accurate determinations based on models built from independent sources.

3.4.2 *Mining from Sparse, Uncertain and Incomplete Data Representation*

Sparse, uncertain, complex and incomplete data are the common defining features for big data applications. If data is sparse then it becomes difficult to take an appropriate decision or reliable conclusion on few data points [23]. This issue arises because of data dimensionality and it becomes difficult to obtain a clear trend or distribution from data. General approaches to resolve these issues are to apply dimension reduction or feature selection to reduce data dimensions by including additional data samples to minimize data scarcity. Dimensional reduction techniques carefully include more additional samples to improve the data scarcity using generic unsupervised machine learning methods and pattern recognition approaches in data mining and big data analysis [43, 44].

In case, data is uncertain then it represents inaccurate data readings and collections where data fields are no longer deterministic and represents some random/error distributions. General approaches to resolve data uncertainty issue are error-aware data mining, Naïve Bayes model, etc.

Incomplete data refers to the missing of data field values for some samples. This scenario takes place because of failure or malfunctioning of a sensor node or system involved in generation of data. However, modern data mining techniques are efficient and have in-built solution like data imputation to handle with these missing values and produce more improved models.

3.5 Mathematical Modelling and Algorithms

In this section, we have discussed big data-based analytics and prediction techniques. In order to provide useful insight to the organizational data and to gain correct content, data must be processed with advanced algorithms and available tools (e.g. analytics and algorithms) for generation of meaningful information from huge data and clusters, and further classify this obtained information or data into various classes based on various interests and meaningful information provided by different customers [39, 45]. These algorithms can be divided into following subgroups.

3.5.1 Probabilistic Learning Model and Statistical Analysis

As stated earlier, uncertain data represents the major challenge in which each data item can be considered as a sample distribution and not like a single value. Therefore, the common solution is to estimate model parameters by taking data distribution into consideration. For example, error-aware data mining [40] can be used to build Naïve Bayes model by finding the mean and the variance values on each single data item. Whereas, in incomplete data, missing value can take place because of many reasons like malfunctioning of the sensor node, use of systematic policies for collection of values after regular interval of time, etc. [42, 44, 46]. Modern computing algorithms are designed in such a way that they can handle missing values efficiently. Data imputation [42, 44, 47] is also another approach designed to handle missing values and to produce improved models.

3.5.2 Fuzzy Rule-Based Expert Systems

It is considered that both model performance and interpretability are of major importance, and require efforts to keep the rule base small and comprehensible. Therefore, Computational Intelligence (CI) approaches for data mining and analysis of big data have been developed for feature selection, feature extraction, model optimization and model reduction [45, 48]. For analysis of massive amount of data, fuzzy logic-based approaches, probabilistic reasoning based, neural networks technique and a set of evolutionary algorithms play a vital role during processing and analysis of big data. These are considered as major components of CI. Each of these algorithms caters the big data committees with complementary reasoning and searching approaches to solve complex and real-world problems.

The selection of a classifier-based model system is defined as the construction of the mathematical model. The mathematical model predicts that a given pattern,

$x_k = [x_{1k}, \ldots, x_{nk}]$, in which $x_k = f_{c1}, \ldots, f_{Cg}$ class should be classified. The classic approach for this problem with C classes is based on Bayes' rule [40, 46, 49].

3.5.3 *Rule Base Reduction Techniques*

Mostly, algorithms are used to obtain suitable classifiers which are based on accuracy or interpretability. In recent study, some algorithms for having been reported which are designed on combining the use of these properties; fuzzy clustering is one of the approaches that can derive transparent models [13], linguistic constraints is another approach that can be are applied to fuzzy modelling and rule extraction from neural networks [9, 13, 47].

3.5.4 *Recommendation Mining*

Recommendation mining-based systems consider input from users' behaviour and from the provided input data. Recommendation systems try to make possible predictions about customer behaviours, their likes and dislikes, and how much attention they have given for each attractive class of advertisements [23, 48].

3.6 Clustering Algorithms

The clustering algorithms are used to design the different clusters of the user or customer interest's on the basis of provided inputs to the recommendation systems or recommendation mining systems [23].

The clustering is a simple approach for grouping a set of objects into classes of similar objects or items. In general, the definition of similarity varies from one clustering model to another. In most of the cases of these models, the concept of similarity has been measured based on various distances such as Euclidean distance or cosine distance. The major objective of data clustering techniques is to arrange a set of n objects into k clusters such that objects in the same cluster are cohesive in nature than objects in different clusters. Therefore, clustering is one of the most popular tools for data exploration and data organization [49]. There are following basic requirements for clustering algorithm to assign any group of similar data items or objects:

- An algorithm
- Similarity and dissimilarity criterions
- A stopping condition

There are various types of the clustering algorithm in data mining that can be used for predictive computing, few of them are explained below.

3.6.1 k-Means Clustering

This clustering technique is categorized under unsupervised learning algorithms. Here, dataset can be classified into different clusters, say k clusters. The major objective is to define or find k centers randomly. A method for finding these k centers might be ad hoc or cunning in nature. It is because by changing the location of k centers will impact on result accordingly. That's why these k centers must be placed far away from each other [49]. In further step, a data point is selected from the dataset and associate to the nearest center. When no point is left, and an early grouping is done [23], k new centroids must be recalculated from the resulting cluster of the previous step. Finally, the squared error function represented by using this algorithm (Eq. 3.5):

$$J(V) = \sum_{i=1}^{c} \sum_{j=1}^{c_i} \left\| x_i - v_j \right\|^2 \tag{3.5}$$

where '$\|x_i - v_j\|$' represents the Euclidean distance, c_i represents the number of data points in ith cluster and c represents the number of cluster centers.

3.6.2 Centroid Generation Using Canopy Clustering

It is also an unsupervised clustering algorithm, often used as preprocessing step for the k-means algorithm or the Hierarchical clustering algorithm. It speeds up the clustering operations on large data sets, whereby the other algorithms may be impractical due to the size of the data sets [50, 51].

3.6.3 Fuzzy k-Means Clustering Technique

As discussed in k-means clustering, on the basis of feature of vector x these clusters can be categorized as 'hard' or 'crisp' clusters. It is because vector x may or may not belong to particular cluster, whereas in fuzzy k-means these clusters are categorized into 'soft' or 'fuzzy' clusters, where feature vector x can have a degree of membership in each cluster [52].

3.7 Classification Algorithms

The classification algorithms classify various provided input to the recommendation systems that learn from existing labelled or categorized documents. The working of classification is based on supervised learning approaches and classifies what documents of a specific category look alike and can assign unlabelled documents to the correct category [52].

The main objective of classification approaches is to classify the labelled unseen documents; therefore, the classification approaches perform the work for grouping the different items with same labelled and together. The classification approaches are the basic machine learning approaches to provide an efficient way for computers to make decisions based on experience and, in the process, emulate certain forms of human decision making. The classification techniques are divided into two groups (1) supervised classification technique and (2) unsupervised classification technique [52]. Table 3.1 illustrates the differences between supervised learning and unsupervised learning-based approaches for the classification.

Various classification algorithms are:

1. Naïve Bayesian [52–54]
2. Complementary Naïve Bayesian [53]
3. Stochastic Gradient Descent (SDG) [55]
4. Random Forest [53]
5. Support Vector Machine [56]

3.8 Summary

In this chapter, we have illustrated the use of predictive computing techniques for predicting the outputs of the massive amount of data. We have described the predictive models for e-Transportation and for the banking sector. e-Transportation is sufficient to predict the shortest route with less traffic congestion on roads. In the current state-of-the-art methods, different organizations in every industry are developing new sensing devices to sense massive big data, as well as to develop analytic computational models and platforms that can help to synthesize the traditional structured data with semi-structured and unstructured sources of information. When

Table 3.1 Differences between supervised and unsupervised machine learning-based approaches

Supervised learning	Unsupervised learning
– Use training data to infer model – Apply model to test data – e.g. Maximum likelihood, perceptron, SVM [52]	– No training data – Model inference and application – Both rely on test data exclusively – e.g. k-means

properly used, predictive computing can provide unique insights into market trends, equipment failures, buying patterns, healthcare, maintenance cycles and many other business issues, lowering costs and enabling more targeted business decisions.

References

1. McAfee A, Brynjolfsson E, Davenport TH, Patil DJ, Barton D (2012) Big data: the management revolution. Harvard Bus Rev 90(10):61–67
2. Jatrniko W, Arsa DMS, Wisesa H, Jati G, Ma'sum MA (2016) A review of big data analytics in the biomedical field. In: Proceedings of international workshop on big data and information security (IWBIS), IEEE, pp 31–41
3. Bezdek JC, Dunn JC (1975) Optimal fuzzy partitions: a heuristic for estimating the parameters in a mixture of normal distributions. IEEE Trans Comput 100(8):835–838
4. Xia F, Wang W, Bekele TM, Liu H (2017) Big scholarly data: a survey. IEEE Trans Big Data 3(1):18–35
5. Ansolabehere S, Hersh E (2012) Validation: what big data reveal about survey misreporting and the real electorate. Polit Anal 20(4):437–459
6. Lynch C (2008) Big data: how do your data grow? Nature 455(7209):28–29
7. Song H, Basanta-Val P, Steed A, Jo M (2017) Next-generation big data analytics: state of the art, challenges, and future research topics. IEEE Trans Ind Inform. doi:10.1109/TII.2017. 2650204
8. Wang S, Bonomi L, Dai W, Chen F, Cheung C, Bloss CS, Cheng S, Jiang X, (2016) Big data privacy in biomedical research. IEEE Trans.Big Data, doi:10.1109/TBDATA.2016.2608848
9. Chen CP, Zhang CY (2014) Data-intensive applications, challenges, techniques and technologies: a survey on big data. Inf Sci 275:314–347
10. Agrawal D, Das S, El Abbadi A (2011) Big data and cloud computing: current state and future opportunities. In: Proceedings of the 14th international conference on extending database technology, ACM, pp 530–533
11. Hashem IAT, Yaqoob I, Anuar NB, Mokhtar S, Gani A, Khan SU (2015) The rise of "big data" on cloud computing: review and open research issues. Inform Syst 47:98–115
12. Fan W, Bifet A (2013) Mining big data: current status, and forecast to the future. ACM SIGKDD Explor Newsl 14(2):1–5
13. Ju H, Hong CS, Takano M, Yoo JH, Chang KY, Yoshihara K, Jeng JY (2013) Management in the Big Data & IoT Era: a report on APNOMS 2012. J Netw Syst Manage 21(3):517–524
14. Xu B, Da Xu L, Cai H, Xie C, Hu J, Bu F (2014) Ubiquitous data accessing method in IoT-based information system for emergency medical services. IEEE Trans Ind Inform 10 (2):1578–1586
15. Chen M, Mao S, Zhang Y, Leung VC (2014) Big data: related technologies, challenges and future prospects. Springer International Publishing, pp 2–9
16. Malekian R, Kavishe AF, Maharaj BTJ, Gupta PK, Singh G, Waschefort H (2016) Smart vehicle navigation system using Hidden Markov Model and RFID sensors. Wirel Pers Commun 90(4):1717–1742
17. Gudivada VN, Baeza-Yates RA, Raghavan VV (2015) Big data: promises and problems. IEEE Comput 48(3):20–23
18. Bonomi F, Milito R, Zhu J, Addepalli S (2012) Fog computing and its role in the internet of things. In: Proceedings of first edition of the MCC workshop on mobile cloud computing, ACM, pp 13–16
19. Gupta PK, Maharaj BTJ, Malekian R (2016) A novel and secure IoT based cloud centric architecture to perform predictive analysis of users activities in sustainable health centers. J Multimedia Tools Appl. doi:10.1007/s11042-016-4050-6

20. Kumar S, Datta D, Singh SK (2015) Black hole algorithm and its applications. In: proceedings of computational intelligence applications in modeling and control, Springer International Publishing, pp 147–170
21. Khan N, Yaqoob I, Hashem IAT, Inayat Z, Mahmoud Ali WK, Alam M, Shiraz M, Gani A (2014) Big data: survey, technologies, opportunities, and challenges. Sci World J. http://dx.doi.org/10.1155/2014/712826
22. Singh D, Reddy CK (2015) A survey on platforms for big data analytics. J Big Data 2(1):8
23. Wu X, Zhu X, Wu GQ, Ding W (2014) Data mining with big data. IEEE Trans knowl Data Eng 26(1):97–107
24. Gu K, Tao D, Qiao JF, Lin W (2017) Learning a no-reference quality assessment model of enhanced images with big data. IEEE Trans Neural Networks Learn Syst. doi:10.1109/TNNLS.2017.2649101
25. Dorina P, Dominic S (2015) Sustainable urban transport in the developing world: beyond megacities. Sustainability 7(7):7784–7805
26. Subrata M (2012) Operational efficiency of freight transportation by road in India. TCI-IIMC Joint study report. Available via TCI. https://www.tcil.com/tcil/pdf/study-report/a-joint-study-report-by-tci&-iim-2009-10.pdf. Accessed 15 May 2017
27. Pattanaik V, Mayank S, Gupta PK, Singh SK (2016) Smart real-time traffic congestion estimation and clustering technique for urban vehicular roads. In: Proceedings of IEEE region 10 conference TENCON, Singapore, pp 3420–3423
28. Kuang L, Hao F, Yang LT, Lin M, Luo C, Min G (2014) A tensor-based approach for big data representation and dimensionality reduction. IEEE Trans Emerg Top Comput 2(3):280–291
29. Jain AK (2010) Data clustering: 50 years beyond K-means. Pattern Recogn Lett 31(8):651–666
30. Shapira G, Chen Y (2016) Common pitfalls of benchmarking big data systems. IEEE Trans Serv Comput 9(1):152–160
31. Zhang H, Chen G, Ooi BC, Tan KL, Zhang M (2015) In-memory big data management and processing: a survey. IEEE Trans Knowl Data Eng 27(7):1920–1948
32. Yu HQ, Zhao X, Zhen X, Dong F, Liu E, Clapworthy G (2014) Healthcare-event driven semantic knowledge extraction with hybrid data repository. In: Proceedings of 4th IEEE international conference on innovative computing technology (INTECH), pp 13–18
33. Clifton C, Kantarcioğlu M, Doan A, Schadow G, Vaidya J, Elmagarmid A, Suciu D (2004) Privacy-preserving data integration and sharing. In: Proceedings of the 9th ACM SIGMOD workshop on research issues in data mining and knowledge discovery, ACM, pp 19–26
34. Fujita H (2016) Big data-based clouds health-care and risk predictions based on ensemble classifiers and subjective projection. In: Proceedings of 17th international conference on computational intelligence and informatics (CINTI), IEEE, pp 11–12
35. Butte S, Patil S (2016) Big data and predictive analytics methods for modeling and analysis of semiconductor manufacturing processes. In: Proceedings of workshop on microelectronics and electron devices (WMED), IEEE, pp 1–5
36. Fernández A, del Río S, López V, Bawakid A, del Jesus MJ, Benítez JM, Herrera F (2014) Big data with cloud computing: an insight on the computing environment, mapreduce, and programming frameworks. Wiley Interdisc Rev Data Min Knowl Discov 4(5):380–409
37. Da Xu L, He W, Li S (2014) Internet of things in industries: a survey. IEEE Trans Ind inform 10(4):2233–2243
38. Wu D, Birge JR (2016) Risk intelligence in big data era: a review and introduction to special issue. IEEE Trans Cybern 46(8):1718–1720
39. Gelman A, Carlin JB, Stern HS, Rubin DB (2014) Bayesian data analysis. Chapman & Hall/CRC, Boca Raton
40. Rice J (2006) Mathematical statistics and data analysis. Nelson education, 3rd edn. Cengage Learning
41. Akthar N, Ahamad MV, Ahmad S (2016) MapReduce model of improved k-means clustering algorithm using hadoop mapReduce. In: Proceedings of 2nd international conference on computational intelligence & communication technology (CICT), IEEE, pp 192–198

42. Chen H, Chiang RH, Storey VC (2012) Business intelligence and analytics: from big data to big impact. MIS Q 36(4):1165–1188
43. Aggarwal CC, Philip SY (2009) A survey of uncertain data algorithms and applications. IEEE Trans Knowl Data Eng 21(5):609–623
44. Kumar S, Sadhya D, Singh D, Singh SK (2015) Cloud security using face recognition. In: Deka G and Bakshi S (eds) Handbook of research on securing cloud-based databases with biometric applications IGI Global, Hershey, pp 298–319. doi:10.4018/978-1-4666-6559-0.ch014
45. López V, del Río S, Benítez JM, Herrera F (2015) Cost-sensitive linguistic fuzzy rule based classification systems under the MapReduce framework for imbalanced big data. Fuzzy Sets Syst 258:5–38
46. Arel I, Rose DC, Karnowski TP (2010) Deep machine learning—a new frontier in artificial intelligence research. IEEE Comput Intell Mag 5(4):13–18
47. Kotsiantis S, Kanellopoulos D (2006) Association rules mining: a recent overview. GESTS Int Trans Comput Sci Eng 32(1):71–82
48. Dean J (2014) Big data, data mining, and machine learning: value creation for business leaders and practitioners. Wiley, New York
49. Hartigan JA, Wong MA (1979) Algorithm AS 136: a k-means clustering algorithm. J Roy Stat Soc 28(1):100–108
50. Cai X, Nie F, Huang H (2013) Multi-view K-means clustering on big data. In: Proceedings of the twenty-third international joint conference on artificial intelligence (IJCAI), pp 2598–2604
51. Shim K (2012) MapReduce algorithms for big data analysis. Proc VLDB Endow 5(12):2016–2017
52. Jain AK, Murty MN, Flynn PJ (1999) Data clustering: a review. ACM comput surv(CSUR) 31(3):264–323
53. Breiman L (2001) Random forests. Mach Learn 45(1):5–32
54. Saad ZS, Glen DR, Chen G, Beauchamp MS, Desai R, Cox RW (2009) A new method for improving functional-to-structural MRI alignment using local Pearson correlation. Neuroimage 44(3):839–848
55. Bottou L (2010) Large-scale machine learning with stochastic gradient descent. In: Proceedings of COMPSTAT'2010, Physica-Verlag, HD, pp 177–186
56. Cortes C, Vapnik V (1995) Support-vector networks. Mach Learn 20(3):273–297. doi:10.1007/BF00994018

Chapter 4
Cloud-Based Predictive Computing

4.1 Introduction

Recent advancements in the field of wireless and multimedia technology as well as computing, hold out promise to perform real-time communication in the various user-based sectors of health, transportation, media and education. The vision of the IoT and cloud computing is one such and provides real-time information to the connected users in the network. This advancement in technologies has led to the issue of data proliferation which in turn is responsible for data explosion and high cost of data processing. This exponential increase in data includes human data on the social media in the form of emails, photos, messages, blogs, tweets, digital data generated by sensors, such as GPS, the business data, the classified data to name a few. It is becoming difficult to store, query, analyze and share the data as the available data are huge in volume and highly complex due to the number of data sources and their interrelationships. The scope of the IoT-based cloud computing is significantly broad and includes the living and non-living entities connected to each other in the network. Making use of predictive computing over this large volume of data collected from various sensors nodes and stored in a cloud environment can be made more scalable, pervasive and easier to deploy using these advanced technologies. The cloud enables the business analytics to scale out the data easily and quickly, which in turn enables them to analyze data archive to identify the developing trends and leads to better customer satisfaction and profitability. Using cloud for predictive computing will make the computing resources delivered as a service and will provide multi-tenancy and shared resources this will also help in managing the issue of data proliferation. Various areas of opportunities can be achieved by using the cloud for predictive computing. Some of which are pre-packed cloud-based solutions, predictive modelling with the data in the cloud, and flexible compute power among many more. The advantages are scalability, pervasiveness, deployment agility, moving analytics to data, whereas the cons are complexity, privacy and security, regulatory issues and moving data to the cloud.

P.K. Gupta et al., *Predictive Computing and Information Security*,
DOI 10.1007/978-981-10-5107-4_4

Knowledge discovery and decision making are one of the major objectives of predictive computing. Big data computing poses a severe challenge in terms of the necessary hardware and software resources required for decision making. Hence, we look upon cloud technology as it offers a promising solution to this challenge by enabling ubiquitous and scalable provisioning of the computing resources [1]. In this chapter, we have explored the hybrid measure of IoT and cloud computing to predict the health status of a user by analyzing his/her physical activities at sustainable health centre of the smart city. It is considered that equipment in these sustainable health centres are equipped with sensors and continuously store the data related to the user's session in the cloud. This stored data is further utilized by the concerned healthcare professional for predicting the health status and in case, if any severe measures are required then alert is sent accordingly.

4.2 Related Work

Use of predictive computing from healthcare perspective is also increasing with the advancement of technologies. Various healthcare professionals and users are using a variety of ways and techniques for analyzing and predicting the health status. To perform prediction, data is collected from a variety of sensors like embedded sensors, wearable sensors and stored in the cloud for its further analysis and finding of patterns. Zhang et al. [2] also explained the use of the wireless sensor networks in healthcare in the future from a ubiquitous perspective and proposed a 3-tier system architecture for healthcare applications. They have also stated that with the technological advancements, the field of medical informatics has focused more on as well as emphasized the use of the Smartphones over wearable devices. Chen et al. [3] have presented the vision of IoT from the perspective of China and specified that in 'Remote medical monitoring' data can be collected from various sensor-like devices placed on an individual's body and once the data is processed advice can accordingly be given. In [4], Islam et al., have presented various aspects of the IoT-based healthcare technologies and stated that IoT can help any age group and address any disease in an innovative manner. In [5], Yang et al., have designed the home mobile healthcare system for wheelchair users. The proposed architecture utilizes the Smartphone for sending and receiving instructions from the source and sink nodes. In [6], Amendola et al., have used the RFID-based wearable tags to identify the movement of the body parts like arms, legs etc.; in fact, they have used the IoT for monitoring the information collected related to human lifestyles.

4.2.1 Cloud-Based Healthcare Frameworks

With the advancement in technologies, a number of cloud-based healthcare frameworks are being designed. Further, integration of clouds with IoT-and big data-based frameworks could provide several benefits of easy to deploy mechanism over traditional networks, enhanced information security during communication, quick access of records and energy savings over traditional existing healthcare frameworks [7]. In [8], Ghulam et al. have also focused on integration of cloud and IoT to have smart health solution. They have stated that convergence of the IoT and the cloud can render a wide application in daily and social life. It is because, IoT is a set of real-world small devices with limited processing power and storage capacity whereas the cloud can have huge storage capacity and processing power. Integration of IoT and cloud depicted by them is presented in Fig. 4.1. They have also identified various attributes like storage, accessibility, processing, distance, big data and security, where cloud computing can provide milestones achievements over IoT.

In [8], Ghulam et al. have also presented a health monitoring framework that captures voice, temperature, humidity, electrocardiogram of a patient using IoT technologies and at cloud side, main components are authentication manager, data manager, feature extraction server, classification server and storage. For authenticity, authors have embedded watermark into the signal. In [9], Tyagi et al. have proposed a cloud-based conceptual framework for implementing a cloud-centric IoT-based healthcare framework. Authors have built a network of various healthcare entities like patients, doctors, hospitals, etc. and used this network for safe transfer of medical information. In [10], Hossain and Muhammad have focused on real-time health monitoring infrastructure for analyzing patients. They have designed and presented a cloud-based healthIIoT framework to monitor ECG and other healthcare-related data using smart phones. Authors have also implemented the used the watermarking techniques for the security of data. In [11], Kashfia et al. have presented a healthcare framework known as 'Cloud-based MEDical system' (CMED) for developing countries. This CMED system consists of a portable health kit, WSN connection and tablet/Smartphone. CMED framework is shown in Fig. 4.2.

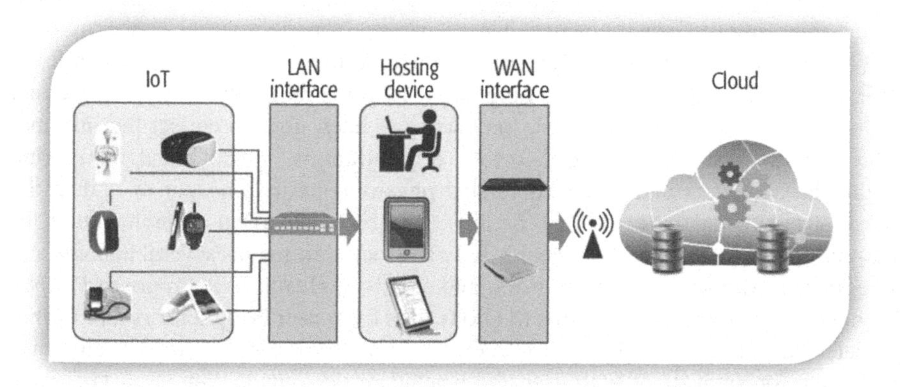

Fig. 4.1 Integration of IoT and cloud [8]

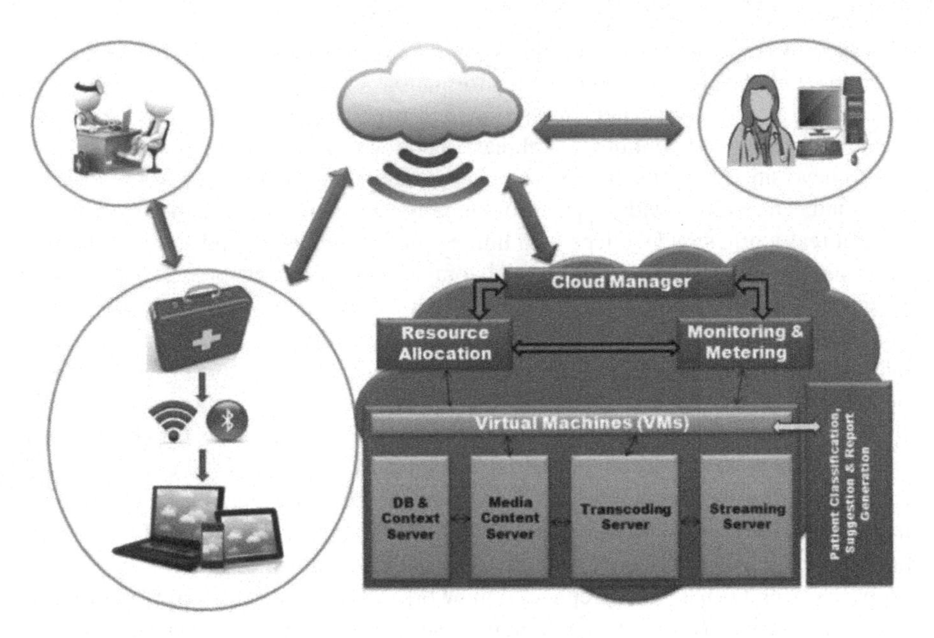

Fig. 4.2 CMED healthcare framework [11]

In [12], Nandyala and Kim have proposed an IoT-based architecture for u-Healthcare monitoring. They have also mentioned that traditional cloud computing architecture faces a lot of challenges and presented an extension of cloud as Fog. They have proposed a Cloud to Fog computing model for implementing the proposed architecture of healthcare monitoring. In [13], Kumar et al., have proposed a RFID-based intelligent authentication scheme in the healthcare using vehicular cloud computing. The authors have placed various tags and readers from users to road side units. Communication between these tags and readers are secured by elliptical curve cryptography-based key generation algorithm. In [14], Hassanalieragh et al., have presented the integration guidelines for remote health monitoring into medicinal practice. They have used smartphones as concentrator in IoT infrastructure and for aggregation of data, cloudlet's or clouds have been used. It is also realized that data processing could be done more efficiently using cloud rather that cloudlet's and wearable sensors have been used to collect the information. Authors have performed 2-3 days of continuous physiological monitoring using these sensors and collected related physiological parameters to update the relevant health database. In [15], Seales et al., have proposed an architecture for content centric networking for Health-IoT. This networking has several benefits and integrates health services and sensors, and clouds in Health-IoT. Proposed PHINet architecture automatically records the body data from user body sensors and further updates the database accordingly. Use of clouds in this architecture makes data easy to handle for analysis of both user and healthcare personnel. Wan et al. [16] have implemented the platform production services using IoT and inter cloud computing architectures. Proposed framework was used by vehicular networking applications. Authors have evaluated the performance of proposed system using probabilistic theory.

4.2.2 Predictive Healthcare Applications

Advances in sensor devices also lead to the development of predictive healthcare applications to be used by e-Health frameworks. Mobile healthcare which is also known as mHealth has only become possible because of this advancement in sensor technology and represents the new opportunities for mHealth-based predictive applications. A wide range of mHealth applications can be found for monitoring of Diabetes [17], Blood Pressure [18], Heart rate [19], Physical activity [20] and Anti-Obesity [21] like areas. There are varieties of healthcare applications available for various types of Smartphone, Tablets and iPad like devices and are known as BlueBox [19], WIHMD [18], AppPoint [22], Heart-To-Go [23], Instant heart rate [24]. Though various types of applications can be find over the internet but somewhere these applications lack in security and privacy, reliability, efficiency and acceptability.

Moreover, these applications are designed for Smartphone like devices where battery life is another issue. Figure 4.3a represents the healthcare application to monitor ECG using Smartphone, Fig. 4.3b shows the mobile application for monitoring of blood Pressure, heart rate, SPO2, Fig. 4.3c represents the capturing of blood pressure monitoring device data using smart phone and Fig. 4.3d represents the information of Glucometer using smartphone. These applications continuously check for the incoming data obtained by sensor for activities like heart rate, ECG, pulse, blood pressure and blood sugar levels of user and keep informing to the healthcare personnel in case of any emergency.

4.3 IoT-Based Cloud-Centric Design Architecture

From the previous discussion, we can observe that most of the proposed healthcare frameworks are over-optimized as they generate large amounts of data, and continuously send alerts to the users and healthcare personnel, which are of no use. Some of these frameworks and applications are designed to use the Smartphone for monitoring and capturing of data. In such a scenario, the continuous connectivity of the Smartphone devices is questionable. Keeping these shortcomings a priority, we have proposed the IoT-based predictive framework as shown in Fig. 4.4 which consists of the evaluation, implementation, feedback and security layers in the IoT environment for a cloud-centric communication between the user and healthcare personnel. IoT-based cloud computing framework architecture is applied for predictive analysis of physical activities of the users in sustainable health centres.

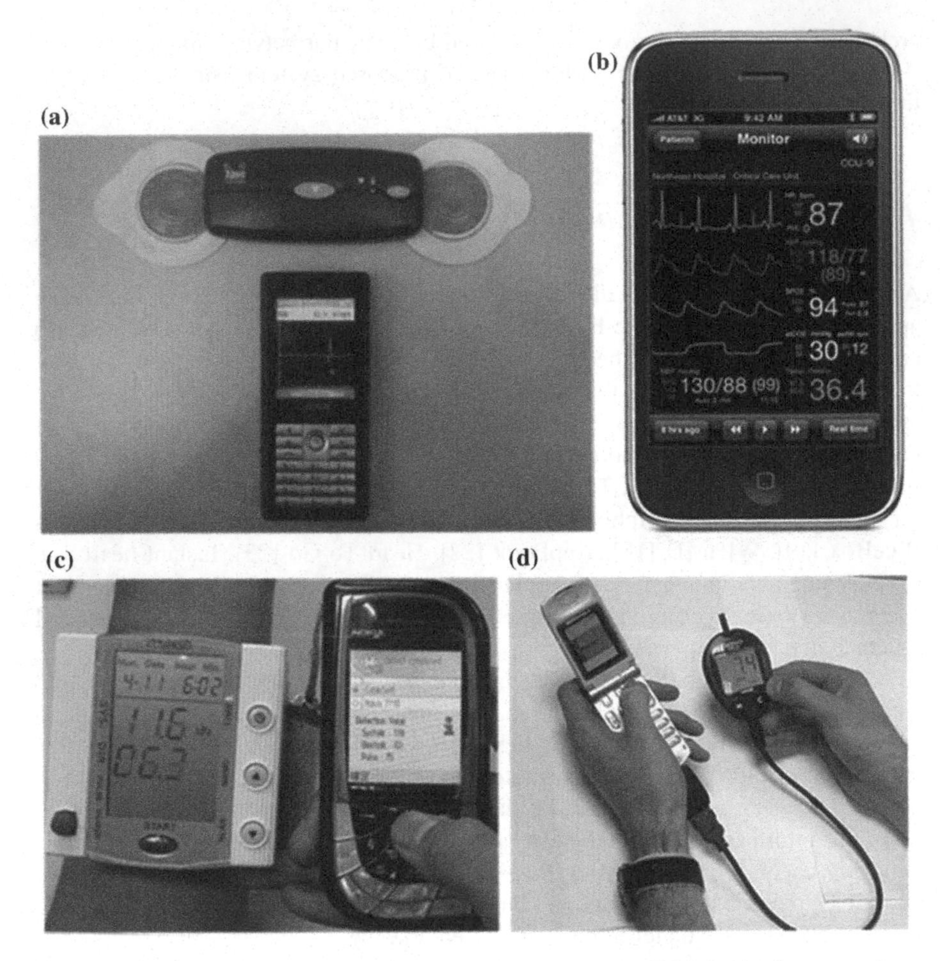

Fig. 4.3 Healthcare applications using Smartphones to monitor **a** ECG, **b** blood pressure, heart rate, SPO2 and body temperature, **c** blood pressure and pulse, **d** blood sugar level

The cloud computing framework provided a system which is embedded with intelligent sensors and devices rather than using smartphone sensors and wearable sensors to sense and retrieve data to store the information (value) of the general health-related parameters for individuals [25]. In this framework, IoT system includes a cloud centre for storing different data, public cloud centre, private cloud data centre and uses the intelligent services for providing the mechanism to secure the stored data and fast communication through intelligent devices. Finally, IoT-based cloud architecture is applied to perform the evaluation for its adoption, prediction analysis of physical activities, efficiency and related security. In the current work, the concept of IoT has been used with the proposed cloud-centric architecture to predict the user's physical activities at sustainable health centre of the smart city. Most healthcare personnel purport that if individual exercises

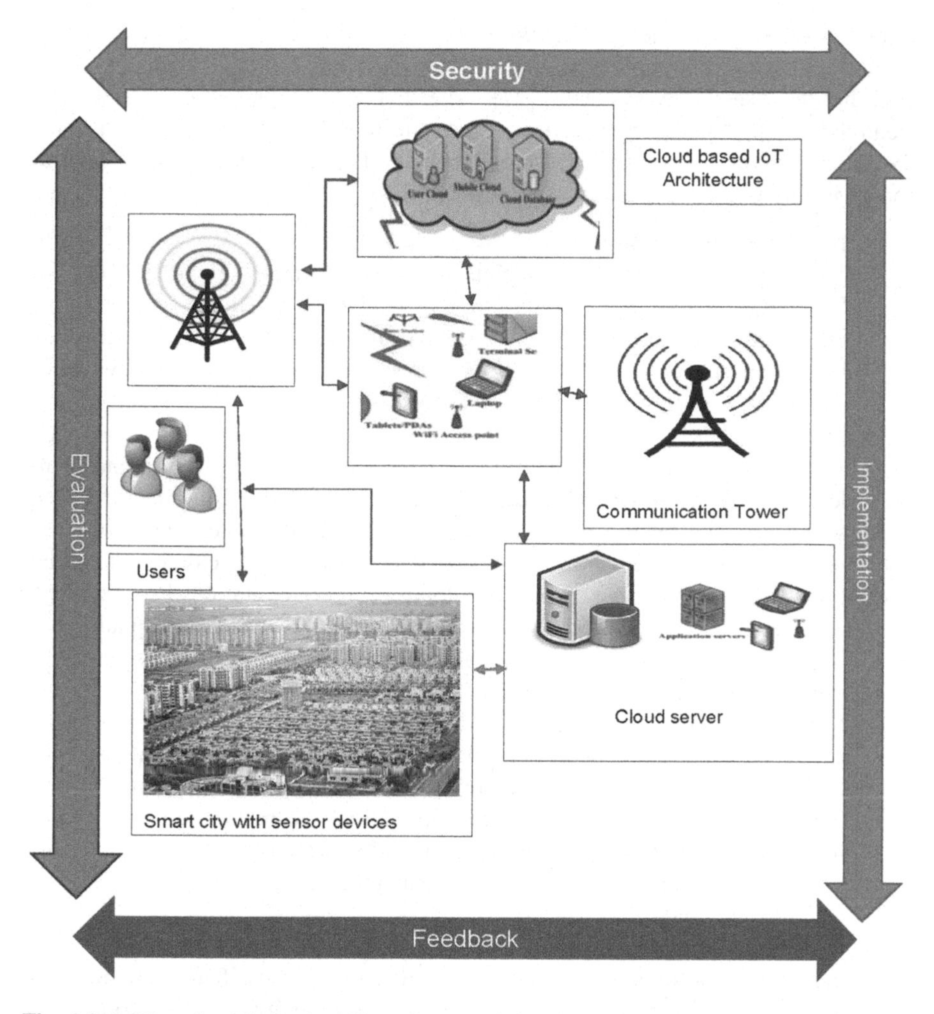

Fig. 4.4 IoT-based predictive healthcare framework for cloud-centric communication

regularly, then the complications and increase in the dosages of medications for diseases like diabetes, blood pressure and heart issues, can get postponed accordingly [26].

In Fig. 4.4, IoT-based cloud computing framework has been developed for providing the solution to individual users using smart devices. The system includes major phases for development of cloud systems. The system consists of the evaluation of sensing data and retrieving information, implementation paradigm, feedback phase and security phase in IoT-based intelligent sensors environment for a cloud-centric communication between the user and healthcare personnel. The basic methodology is applied to implement the IoT-based computing models or framework for the cloud-centric system for providing the efficient solution in the health sectors. The IoT-based cloud framework includes four different phases in the cloud system-based model. These phases are applied to cater the faster data processing and data communication between the individual and IoT-based systems (e.g. health assessment systems) [27]. The IoT-based computing models and methodology also can be applied to mitigate the redundancy and sparsity of massive amount of data which are necessary to be stored for analysis purposes. The four phases in this process are given as follows [26]:

(i) **Implementation phase**
The implementation phase depicts the overall framework architecture of IoT-based computing systems from the data acquisition, extraction of information, to data storage and analysis of data. It also uses the embedded intelligent sensors or devices in compression of any wearable devices or intelligent sensors by the individual.

(ii) **Security phase**
Security phase mainly emphasizes on IoT-related security issues and challenges on data storage and transfer extracted knowledge from data between source and destination intelligent sensors [28]. It also caters the better services for implementation of distributed system and cloud computing framework for various types of cloud servers (cloud database servers, data-centric server, database servers). The main objective of this phase is to implement the better security to user privacy preservation and security to user information.

(iii) **Feedback phase**
In IoT-based cloud computing framework, feedback phase uses the results of the evaluation phase which are applied to revise the different applications. After each cycle of the performed application, the reported component or modalities are improved by doing rectification which assists to improve the overall performance of the IoT-based framework.

(iv) **Evaluation phase**
Evaluation phase is a phase which performs the functions of IoT-based computing and modelling systems by utilization of the different applications and to ensure a mitigation in the design complexity and its adoption to improve different services (e.g., healthcare services, agriculture monitoring services, traffic signal monitoring, transportation services, smart village-based services, medical services and information services between remote systems, etc.) [29].

4.3.1 Application Architecture

Application architecture for the IoT-based system proposed above reduces the complexity of the information storage, communication and enhances the performance of the overall healthcare system. The proposed application architecture is modified form of traditional 3-Tier architecture that represents the client-tier, business logic at middle tier and database at server-side. In the proposed application

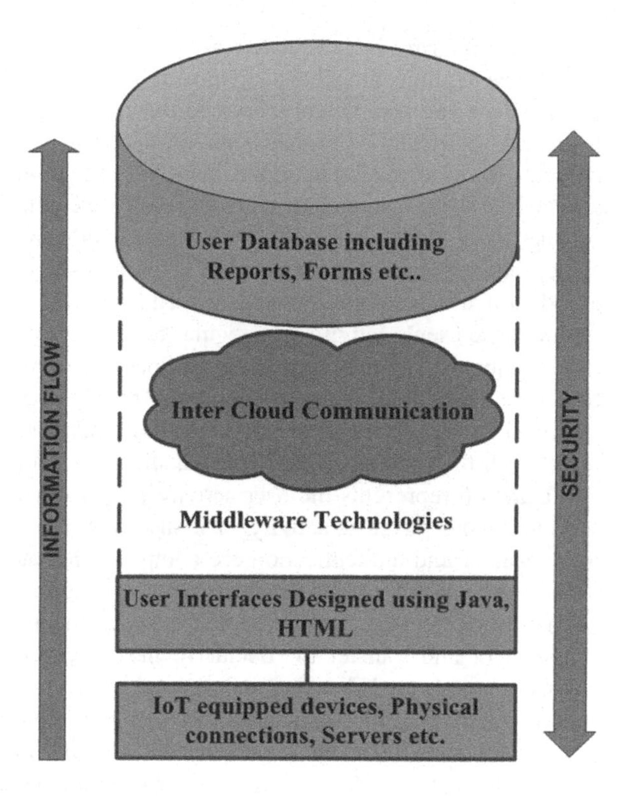

Fig. 4.5 Application architecture

architecture [26], various web services and security services reside inside the cloud which are further implemented as a part of middleware technologies Fig. 4.5 introduces the multilayer application architecture which includes the IoT sensor devices, Treadmill equipment, application servers, base stations and handheld devices like PDAs, Tablets, notebooks, etc. at the bottom. These interconnected devices communicate with the user interfaces designed with JAVA. These user interfaces are responsible for capturing the information from the connected sensors of the treadmill equipment at the end of the user session and any further update of this information with the cloud servers. These user interfaces are further connected via the middleware with many Cloud Servers and support the inter cloud communication with cloud database. In middleware technologies, XML web services can be used to carry secure information from these interfaces to the cloud and can also communicate with the required database for reporting to the user's request.

4.3.2 Predictive Framework for User Activity

Presented predictive user activity framework keeps both the users and healthcare personnel updated [26]. Here, user of the system has been considered as a smart user and performs health-related activities in sustainable health centres. These sustainable health centres have equipment fitted with sensors which are sufficient to capture the basic parameters related to the users' health activities, such as average heart rate, total calories burned, total distance and average speed. This framework captures the information required once the user completes the exercise. Regular activities in these health centres are believed to be able to keep a user fit rather than on the heavy dependence on medicines. Healthcare personnel also suggest the same, and encourage regular physical activity to keep healthy for a longer time and avoid an overdose of the prescribed medicines. The proposed activity framework monitors the user's activity on a regular basis and updates the captured information to the required cloud. In case the user does not perform an activity for a specific time period then this information will also be updated in the cloud database and an alert will be sent to the respective healthcare personnel, who can further communicate with the user regarding understanding the status of his health.

Figure 4.6 represents the user activity framework from the beginning and until the end of the physical activity in a sustainable health center. As the biometric identification and authentication are completed for the user, a secure session gets established with the local database server to monitor the user's activity on that equipment. Here, it is considered that the equipment has embedded sensors which can monitor and transfer the details of the basic parameters like heart rate, total calories burned, total distance and average speed to the connected local database servers. As soon as the physical activity session is over, the equipment stops and the respective values of the basic parameters are stored in the local database server of the sustainable health center. In case the user terminates the session in the middle of the physical activity then that session will be marked INCOMPLETE and the

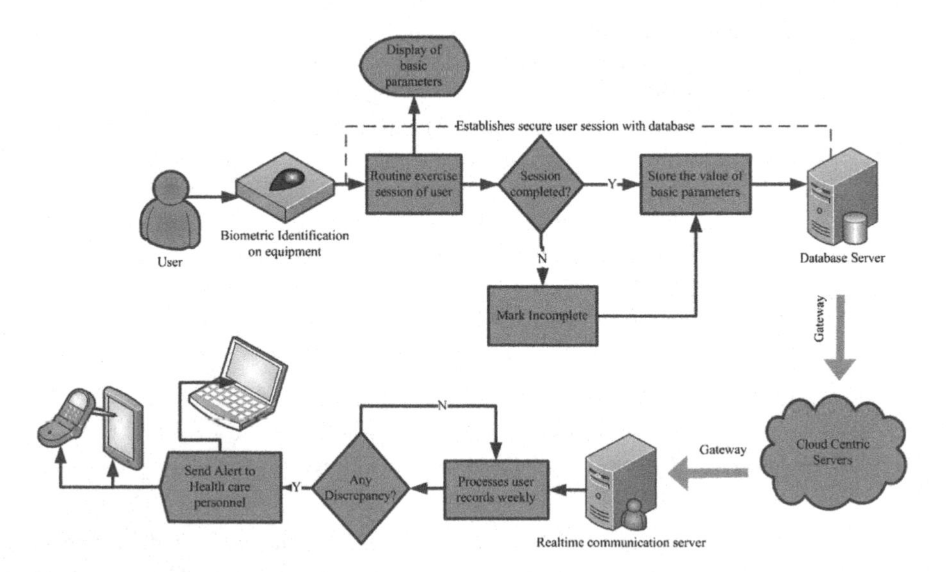

Fig. 4.6 Predictive user activity framework

respective values of the basic parameters are stored in the database. In the next step of this framework, the information stored at the local level is transferred via the gateway to the various cloud servers. These cloud servers keep the information up-to-date and secure. Later on, this stored information is picked up by the real-time communication servers located geographically with the different healthcare personnel. These real-time communication servers process the record thus obtained and relate them to the user on a weekly basis to produce the reports accordingly. If the user regularly terminates the physical activity sessions, before the prescribed activity time for particular equipment or is absent from the sustainable health center then an ALERT message is sent to the user's healthcare personnel either in the form text message or an Email alert.

4.4 Cloud-Based Predictive Computing Design

The proposed cloud-centric architecture reduces the overall complexity of the implementation of the system [26]. It implements various security measures at each step of the communication of the information. As shown in Fig. 4.7, the initial level collects and stores the information for each user from the sustainable health centres into their database server at the local level. The information collected represents the daily values of the basic parameters the specific user has worked out. To communicate this information thus gathered over the cloud the XML web services have

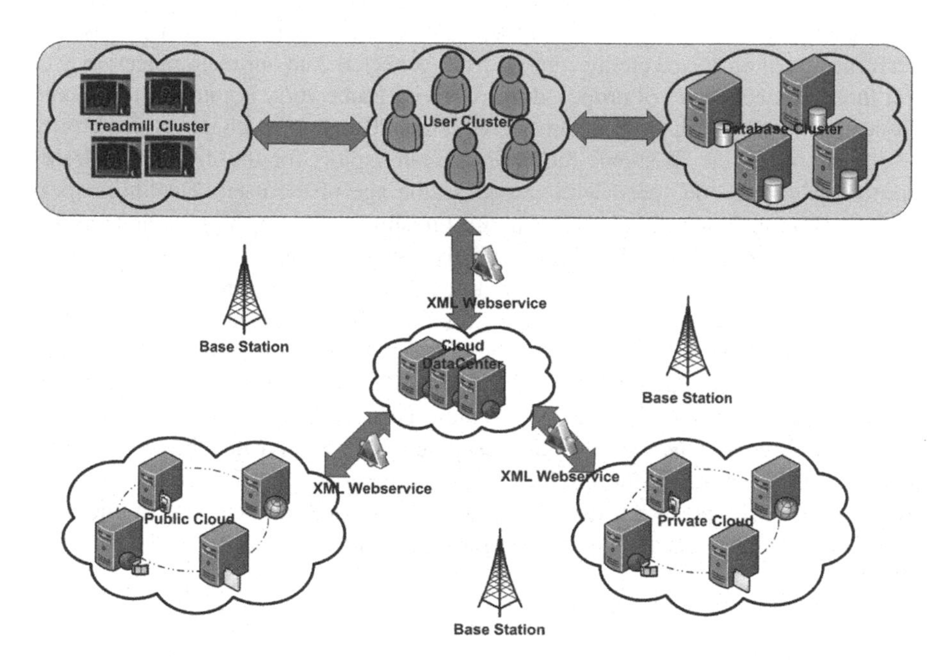

Fig. 4.7 Cloud-based predictive computing design

been used as these are considered secure and faster. From the initial level, the information is collected from the various database clusters distributed to the cloud-centric data center. These data centres also ensure the security of the stored information and further disseminate the information to the public and private clouds.

Both private and public clouds are equipped with a mobile information server, content management server, file server and streaming media servers. These servers continuously exchange information whenever required with their respective clouds. While implementing this architecture, different algorithms are used with different types of clouds to maintain the integrity, consistency and security of the user's information.

4.4.1 Predictive Analysis of Physical Activities

The proposed framework presents the ease of interface to both the healthcare personnel and users. While performing a physical activity in a sustainable health centre, it represents no interface to the user except for identification, authentication and establishing the session with the local database server which is responsible for storing the values of the basic parameters as the user completes the session. We have performed the analysis on average data collected in real time from five different nodes up-to duration of 30 days for time interval 20 min every day. To vary the considered data for experimental purpose requires more nodes and time so that the data could be collected in real time. However, this will not affect the overall performance of proposed architecture as from collected data our main objective is to test the alert mechanism of proposed user activity framework. Figure 4.8 represents the values of the basic parameters stored after 20 min of each physical activity session by the users. It shows the various scatter plots for the distance, calories burned, heart rate and speed with respect to the age of the users. Healthcare personnel can thus define the daily maximum threshold value of these parameters for their users individually and further predictive analysis can be performed accordingly. Here, Fig. 4.8a represents the threshold value for distance ≥ 4.5 km, Fig. 4.8b represents the threshold value for the calories burned ≥ 340, Fig. 4.8c represents the threshold value for the heart rate ≥ 170 bpm, and Fig. 4.8d represents the threshold value for speed ≥ 13 km/h. While performing prediction analysis, the defined threshold values of these basic parameters are checked and if any of value exceeds the limit then an alert is sent to respective healthcare personnel to initiate the necessary action. Also, if a user misses the physical activity session for a whole day or terminates the session in between, then also a message 'INCOMPLETE/ABSENT' gets stored into the database.

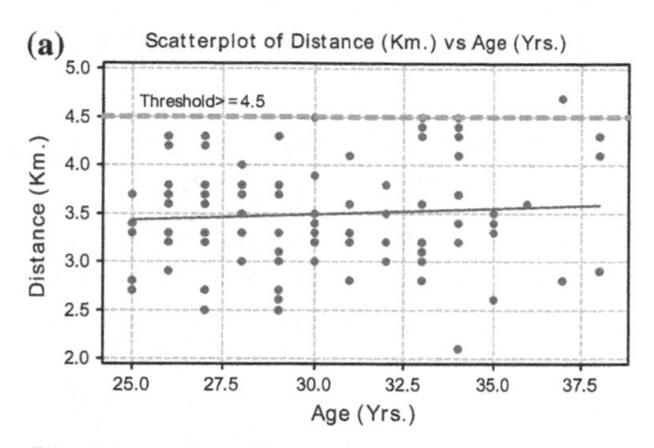

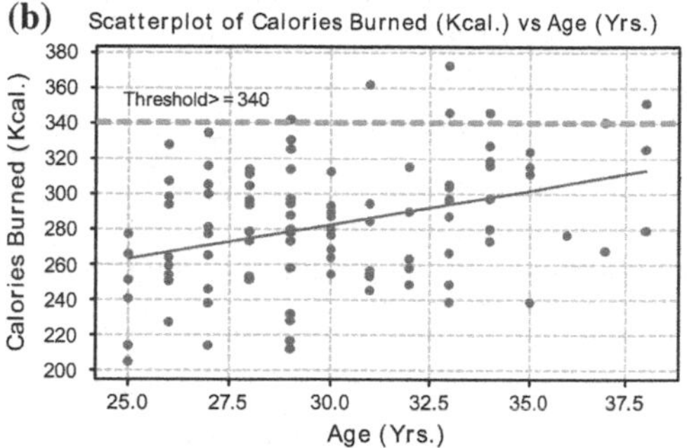

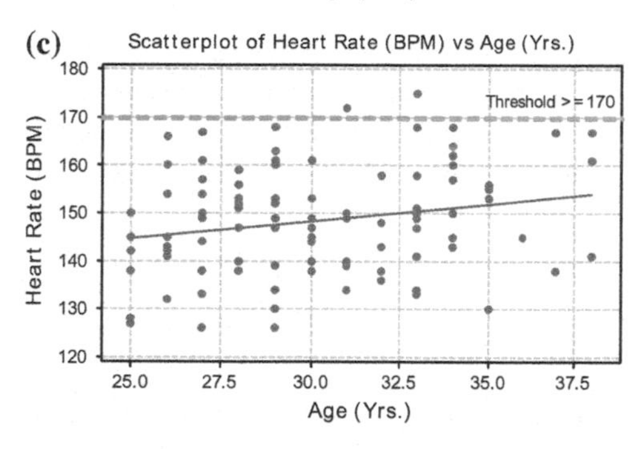

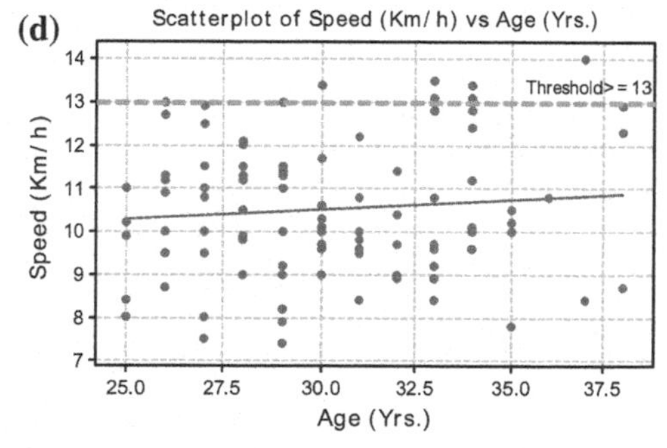

Fig. 4.8 Scatter analysis of users on treadmill **a** distance (km) versus age (years), **b** calories burned (Kcal) versus age (years), **c** heart rate (bpm) versus age (years), **d** speed (km) versus age (years)

4.4.2 *Predictive Efficiency of Framework*

The efficiency of the proposed framework is evaluated using the total communication time, and end-to-end delay.

- *Total communication time* represents the total time taken for the storage of information from the local database server to the cloud data centre and further to the public or private cloud. This time can exceed if there is any alert sent from the public or private cloud to the healthcare personnel. Here, Eqs. 4.1 and 4.2 show the computation of the total transmission time where n represents the number of clusters, t is the time, and w the total offloading iterations and x total number of alerts.

$$T_{\text{TIME}} = \sum_{i=1}^{n} (t(w) + t(x)) \tag{4.1}$$

If there is no alert then the value of $t(x) = 0$ and Eq. 4.1 will become:

$$T_{\text{TIME}} = \sum_{i=1}^{n} (t(w)) \tag{4.2}$$

- *End-to-end delay* represents the information delay time between the nodes. This can be computed by subtracting the minimum time of communication from the total time of communication. Its minimum value shows the early arrival of information at the destination node (Eqs. 4.3 and 4.4).

$$T_{\text{delay}} = T_{\text{TIME}} - T_{\text{minimum}} \tag{4.3}$$

$$T_{\text{minimum}} = f(T_{\text{TIME}}, \text{minimum}) \tag{4.4}$$

Figure 4.9 represents the total transmission time for both the public cloud and private cloud. Here, the value of T_{TIME} reaches up to 12 s for the public cloud and

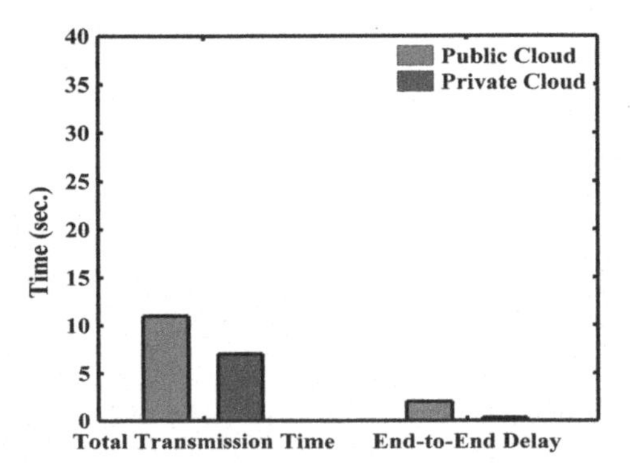

Fig. 4.9 Efficiency of cloud-centric framework

7 s in case of the private cloud. The higher value of T_{TIME} for the public cloud shows that more users are connected with it, as the services are not payable to this cloud.

4.5 Summary

In this chapter, we have presented an integrated view of IoT and cloud computing and proposed a predictive healthcare framework for monitoring of physical activity of user. The proposed framework simplifies several of these issues and stops the proliferation of information. This framework can be used by user of any age group. The proposed framework, has been evaluated on the basis of predictive analysis of physical activities, and its efficiency. There was also an understanding that to minimize the overall cost there should be a minimum time delay between the advice time and issue reported time. This can be observed from the analysis performed for the end-to-end delay for the proposed framework. Predictive analysis of the physical activities represents the results collected in real time using the sensors for the various users on the treadmill for a 20 min time duration. These results also present the common threshold value for all the users, although the healthcare personnel can define it separately for the individual users. Overall, the architecture presented and discussed is more robust and secure in nature, although a lot of work could still be done in the future for its improvement. The issue of load balancing and information distribution throughout the cloud servers can be considered, and proposed framework can be extended to include more health parameters and activities.

References

1. Sharma SK, Wang X (2017) Live data analytics with collaborative edge and cloud processing in wireless IoT networks. IEEE Access 5:4621–4635
2. Zhang Y, Sun L, Song H, Cao X (2014) Ubiquitous wsn for healthcare: recent advances and future prospects. Internet Thing 1(4):311–318
3. Chen S, Xu H, Liu D, Hu B, Wang H (2014) A vision of IoT: applications, challenges, and opportunities with China perspective. Internet Things 1(4):349–359
4. Islam R, Kwak SM, Humaun D, Kabir M, Hossain M, Kwak KS (2015) The internet of things for healthcare: a comprehensive survey. IEEE Access 3:678–708
5. Lin Y, Ge Y, Li W, Rao W, Shen W (2014) A home mobile healthcare system for wheelchair users. In: Proceedings of 18th international conference on computer supported cooperative work in design (CSCWD), IEEE, pp 609–614
6. Amendola S, Lodato R, Manzari S, Occhiuzzi C, Marrocco G (2014) RFID technology for IoT-based personal healthcare in smart spaces. Internet Things 1(2):144–152
7. Bhatt CM, Dey N, Ashour A (2017) Internet of things and big data technologies for next generation healthcare, vol 23. Springer International Publishing, Cham

8. Muhammad G, Rahman SMM, Alelaiwi A, Alamri A (2017) Smart health solution integrating IoT and cloud: a case study of voice pathology monitoring. IEEE Commun Mag 55(1):69–73
9. Tyagi S, Agarwal A, Maheshwari P (2016) A conceptual framework for IoT-based healthcare system using cloud computing. In: Proceedings of 6th international conference-cloud system and big data engineering (Confluence), pp 503–507
10. Hossain MS, Muhammad G (2016) Cloud-assisted industrial internet of things (iiot)? Enabled framework for health monitoring. Comput Netw 101:192–202
11. Sailunaz K, Alhussein M, Shahiduzzaman M, Anowar F, Al Mamun KA (2016) CMED: cloud based medical system framework for rural health monitoring in developing countries. Comput Electr Eng 53:469–481
12. Nandyala CS, Kim HK (2016) From cloud to fog and IoT-based real-time u-healthcare monitoring for smart homes and hospitals. Atlantic 10(2):187–196
13. Kumar N, Kaur K, Misra SC, Iqbal R (2015) An intelligent RFID-enabled authentication scheme for healthcare applications in vehicular mobile cloud. Peer-to-Peer Netw Appl 1–17
14. Hassanalieragh M, Page A, Soyata T, Sharma G, Aktas M, Mateos G, Kantarci B, Andreescu S (2015) Health monitoring and management using internet-of-things (IoT) sensing with cloud based processing: opportunities and challenges. In: Proceedings of IEEE international conference on services computing (SCC), New York, USA, pp 285–292
15. Seales C, Do T, Belyi E, Kumar S (2015) PHINet: A plug-n-play content-centric testbed framework for health-internet of things. In: Proceedings of IEEE international conference on mobile services (MS), New York, USA, pp 368–375
16. Wan J, Zou C, Zhou K, Lu R, Li D (2014) IoT sensing framework with inter-cloud computing capability in vehicular networking. Electron Commer Res 14(3):389–416
17. Franklin V, Greene A, Waller A, Greene S, Pagliari C (2008) Patients engagement with sweet talk—a text messaging support system for young peoplewith diabetes. J Med Internet Res 10 (2):6
18. Kang JM, Yoo T, Kim HC (2006) A wrist-worn integrated health monitoring instrument with a telereporting device for telemedicine and tele care. IEEE Trans Instrum Meas 55(5): 1655–1661
19. Pollonini L, Rajan NO, Xu S, Madala S, Dacso CC (2012) A novel handheld device for use in remote patient monitoring of heart failure patients? Design and preliminary validation on healthy subjects. J Med Syst 36(2):653–659
20. Kwapisz JR, Weiss GM, Moore SA (2011) Activity recognition using cell phone accelerometers. ACM Sig KDD Explor Newsl 12(2):74–82
21. Moghaddam RF, Moghaddam FF, Cheriet M (2011) The bluenetwork concept. University of Quebec, Montreal, Canada, pp 1–8
22. Miloševic M, Shrove MT, Jovanov E (2011) Applications of smartphones for ubiquitous health monitoring and wellbeing management. JITA—J Inf Technol Appl (Banja Luka)-APEIRON 1:1
23. Oresko JJ, Jin Z, Cheng J, Huang S, Sun Y, Duschl H, Cheng AC (2010) A wearable smartphone-based platform for real-time cardiovascular disease detection via electrocardiogram processing. IEEE Trans Inf Technol Biomed 14(3):734–740
24. Azumia Instant heart rate. http://www.azumio.com/s/instantheartrate/index.html. Accessed 15 Apr 2015
25. Ramaswamy L, Lawson V, Gogineni SV (2013) Towards a quality-centric big data architecture for federated sensor services. In: Proceedings of 2013 international congress on big data (Bigdata congress), IEEE, pp 86–93
26. Gupta PK, Maharaj BT, Malekian R (2016) A novel and secure IoT based cloud centric architecture to perform predictive analysis of users activities in sustainable health centres. Multimedia Tools Appl. doi:10.1007/s11042-016-4050-6
27. Fortino G, Trunfio P (2014) Internet of things based on smart objects. Springer International Publishing, Cham

28. Lea R, Blackstock M, (2014) City hub: a cloud-based IoT platform for smart cities. In: Proceedings of 6th international conference on cloud computing technology and science (CloudCom), IEEE, pp 799–804
29. Bonomi F, Milito R, Zhu J, Addepalli S (2012) Fog computing and its role in the internet of things. In: Proceedings of the first edition of the MCC workshop on mobile cloud computing, ACM, pp 13–16

Chapter 5
Internet of Things Based Predictive Computing

5.1 Introduction

The IoT is an emerging research area which is gaining more proliferations due to wide range of applications and uses in different intelligent devices. The connected intelligent devices and extensively interconnected objects are identifiable and equipped with sensing, computing and communication capabilities [1–3]. Therefore, IoT-based computing paradigms provide fast computations and transferring the sensed data onto connected intelligent devices [2, 3]. Atzori et al. [3] defined the '*IoT*' and illustrated those various paradigms for effective computation and transferring of the relevant information between smart devices. Generally, the computing paradigms in the IoT-based frameworks include (1) Internet-oriented computing and communication-based paradigm known as middleware. In the Internet, objects (devices) are connected by various protocols and networking topologies (e.g. star topology, ring topology, mesh topology and other networking topologies) using Internet services, (2) things-oriented based computing paradigms known as intelligent smart sensing devices such as intelligent sensors and actuators. In this paradigm, the intelligent sensing devices are connected to other devices and these devices are capable of receiving and transferring the sensed data. After computation of sensed data, these devices extract relevant information from the captured data and transfer the retrieved information to the other devices, and (3) Semantic-oriented based computing paradigms, known as sharing knowledge-based computation models [4, 5]. In semantic-oriented computing based paradigm, the retrieved information of captured data is computed by predictive computation techniques and relevant information is transferred among connected smart devices and sensors by means of semantic structures of connected devices and its standard. This needs appropriate techniques for retrieving information from the data. The advantages of IoT-based computing paradigms can be unleashed only in the application domain where the three models intersect as mentioned above. IoT is a computing paradigm for smart sensors and devices. The propagation of these

P.K. Gupta et al., *Predictive Computing and Information Security*,
DOI 10.1007/978-981-10-5107-4_5

smart devices builds the IoT in the communication-based network, wherein sensors and other physical devices combine seamlessly with the environment around us, and the information is shared and distributed across different computing platforms in order to develop a common IoT-based framework.

In the IoT scenario, computing framework allows several intelligence sensors and connected devices to cater quick services and fast sharing of knowledge across different platforms [6, 7]. During sharing of knowledge, first, it considers the pervasive presence in the environment of different things or objects that are connected through Internet communication (e.g. wireless and wired communication network). For connection and addressing for devices, distributions of unique addressing schemes are used to provide the standard paradigms to interact the intelligent devices with other devices. These computing platforms cooperate with other things/objects to achieve the common objective by creating new applications and efficient services [8]. The common objectives of IoT-based computing framework can be obtained by integrating sensor–actuator–Internet frameworks and technological aspects. This platform enables things or objects to be connected anytime, anyplace, with anything and anyone using any connected or non-connected path or any networking protocols and any service. The deployment of various sensor based devices in the IoT-framework can be used to provide the various facilities to the different consumers. For enabling the privacy protection and security concern for the individual, IoT-based frameworks are also used in the health care. In the health care, the wearable sensor devices with smart sensors, IoT-based computing system has been envisioned for the past few decades. IoT provides a computing platform to accomplish this vision using body area network, smart devices and sensors. The IoT-based frameworks back end to upload the sensed data by connected devices to servers. In the transportation field, the traffic management and infrastructure monitoring, intelligent transportation and path optimization are also deployed based on real-time traffic information using IoT-based system. The deployed intelligent sensors and physical devices are connected to build an IoT-based infrastructure for monitoring the structural fatigue and other maintenance issues. It also provides a good platform for monitoring of accidents.

5.2 IoT Applications

In this section, various application domains are discussed in detail which gained more proliferation due to IoT. These diverse applications are categorized based on various types of resources availability. The resources include network availability, coverage, scale up and down, heterogeneity of data, repeatability of the different processes, the involvement of customer and their impact [9]. The applications of IOT are shown in Fig. 5.1. Figure 5.2 shows an overview of services by Internet of Things.

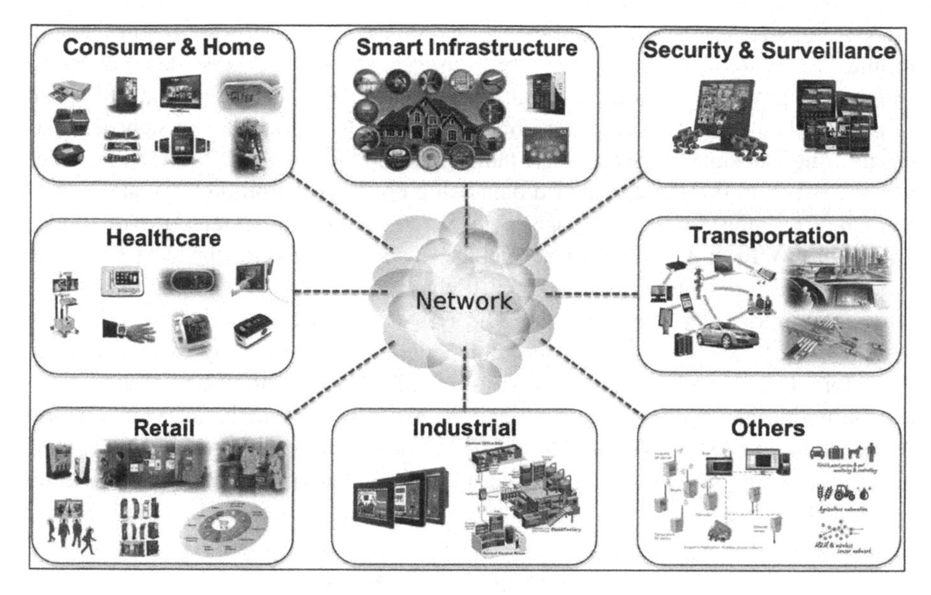

Fig. 5.1 Overview of Internet of Things based applications

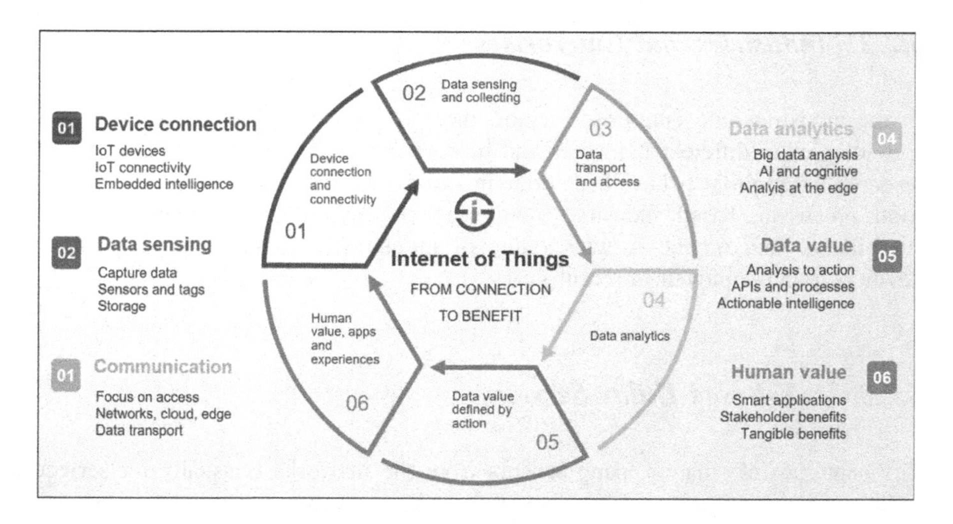

Fig. 5.2 Overview of Internet of Things services

5.2.1 Personal and Home Services

The collected information of sensors such as room temperature by temperature sensors, humidity data, etc., is used for various applications. This information is used by an individual person who directly owns the network. Generally, Internetworking is deployed as the backbone enabling higher bandwidth data (video) transfer as well as to get the higher sampling rates. In the field of health monitoring, ubiquitous sensors have been used for the last two decades. IoT-based framework provides a good platform to realize the home services using different sensors and to upload the captured or sensed data to servers. For instance, smart devices such as smart phones, smart tablets and intelligent sensors can be used for sensing data and transfer the sensed data using communication protocol along with several communication interfaces. The communication interface includes Bluetooth, WiFi and other for interfacing sensors to measure the data based on physiological parameters. The personal body area network is also gaining importance to monitor elder's care at home. The system provides services to an individual at home. It allows the doctor to monitor patients and the elderly in their homes. Therefore, home monitoring system plays a vital role to provide efficient services to the individual by reducing hospitalization costs through early intervention and better treatment to individuals [9, 10]. The air conditioners, washing machines, automatic ovens, refrigerators and other electronic equipments can be joined with intelligent sensors to control home and allow better home facility and energy management. The consumers and individual become fully involved and connected in the IoT revolution in the same manner as the Internet revolution itself [11–13].

5.2.2 Industries and Enterprises

In the industries and enterprises sector, there is a growing interest in using IoT technologies in different industries and businesses [4]. Many industrial IoT-based research-based projects have been done in various areas such as health, agriculture, food processing-based industry, monitoring of environmental effects, security surveillance and others. A wide range of industrial IoT applications has been developed and deployed in recent years.

5.2.3 IoT-Based Utility Services

The captured information using sensors from the networks is usually for service utility and optimization of services rather than consumer consumption in the application domains. It is used by different utility companies. These consist of wide range of networks for regional and national scales. These networks are used for critical utilities monitoring and efficient resource management. The wide range of networks used between cellular communication, WiFi communication and satellite communication. In the current scenario, smart grid-based communication frameworks and smart metering-based communication frameworks are also used for potential applications of the IoT. These are being implemented throughout the world [14]. Efficient consumption of energy is also achieved by proper monitoring of electricity point within the house and using this information to modify the way electricity is consumed. This information at the city scale is used for maintaining the load balance within the grid ensuring high quality of service.

5.2.4 Internet of Things in Healthcare

IoT has received much attention in the healthcare sector. The healthcare services using IoT raises a powerful research domain where the embedded intelligent sensors and physical devices are connected and these sensors and devices are capable of exchanging the information over the network. In these systems, the remote health monitoring of individual includes complete functionalities for long-term diagnosis and recording of health parameters, and detection of critical diseases and health monitoring based on recorded physiological information of the patient [4]. Based on available literature, the healthcare monitoring and diagnosis-based IoT frameworks consist of architecture of three levels: (1) body sensor network that includes body-wearable intelligent sensors for the recording of data from the different part of the body. It works as a unit for data acquisition from an individual body. The acquisition of data includes blood pressure, heart status and body temperature, ECG signals, (2) the second tier consists of complete set-up of communication, networking and the services-based framework. The integrated framework is capable of collecting the sensed data from equipped intelligent sensors from the human body. The collected information is forwarded to the next level [5, 6]. (3) The final third level generally includes the processing, normalization of collected data and analysing the data for retrieving the information for monitoring of health. Figure 5.3 depicts the working architecture of healthcare system using IoT [7, 8].

The IoT-based cloud-centric architectures are also applied for deployment of e-Health based predictive analysis of physical activities of the users in sustainable health centres.

Figure 5.4 depicts a framework for monitoring of individual health. In IoT-based computing paradigm, the predictive model uses the mathematical approaches and computational techniques to predict an event or outcome in the connected or shared smart objects. In this context, the multidisciplinary researchers, data scientists and engineers have begun to design and develop models, and IoT-based computing

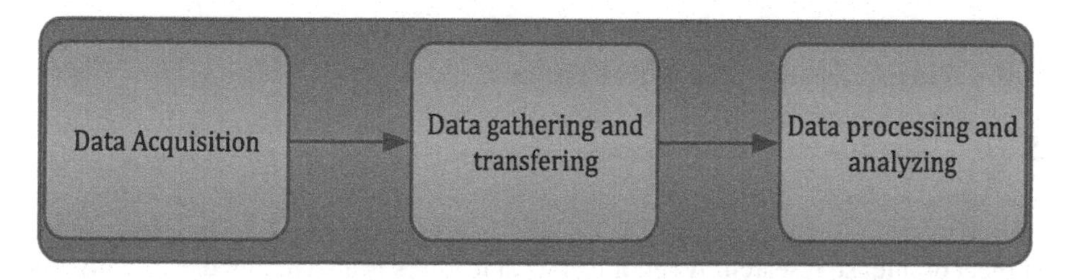

Fig. 5.3 Healthcare monitoring system architecture

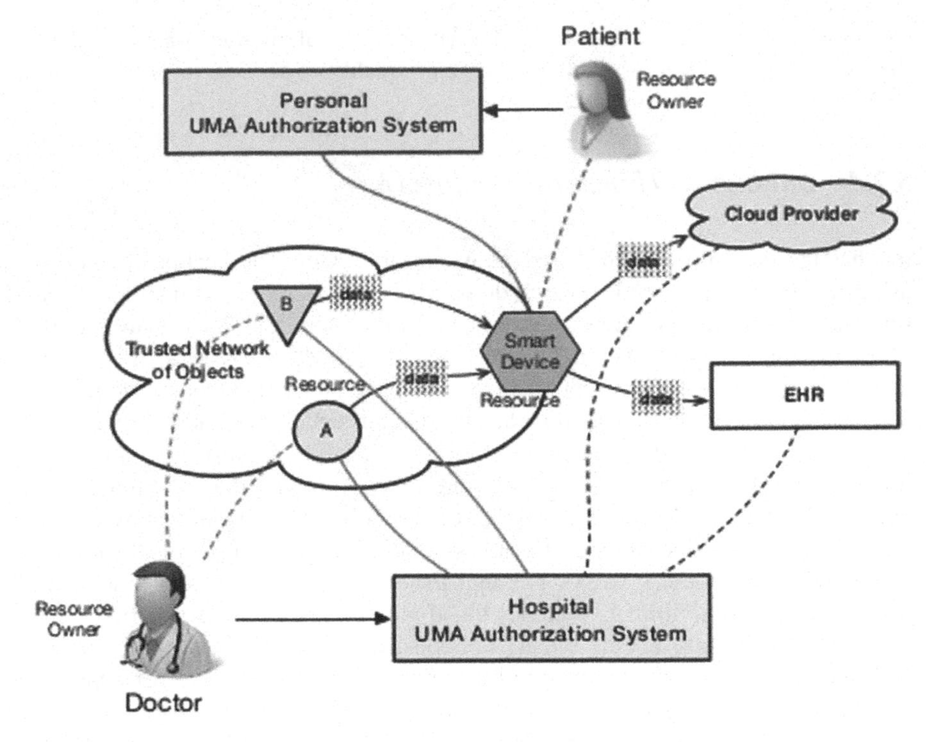

Fig. 5.4 Block diagram of IoT-based framework for health centre

systems to solve the major challenges in using the predictive analytics techniques and computing approaches [8, 15]. These techniques provide the better platform to create an intelligent and smart world, which is equipped with smart sensors and other devices. These connected devices and smart sensors makes possible to convert the real, digital and the virtual platforms into energy efficient smart environment, and provides faster means of communication and transportation, for developing smart cities and other areas [16].

The different devices are currently available commercially for different application-oriented uses and purposes including personal healthcare, activity awareness and fitness.

5.3 Major Issues and Challenges

5.3.1 Technical Challenges

The technical challenges of the IoT-based intelligent devices are numerous and subject of intense research. A set of technical features is also required especially for industrial applications, depending on different applications [17]. These applications include comprehensive capabilities for sensing, and communications in terms of types of intelligent sensors and high sampling rate of received data, fast communication through wireless communication and Internet protocols, data transferring mechanism between intelligent sensors and receivers and computation of precise time for the synchronous collection of data both in one-hop and more hop Internet networks. Moreover, device packaging and defined protocols of intelligent sensors also play a vital role in the proper utilization of the extracted features, and transferring of the massive amount of the data. It is also required for the industrial application, medical science applications, prediction of unknown data based on discriminatory information of extracted features [18, 19]. These intelligent sensors, physically connected devices and their computation capability need the predictive computation based modelling system for forecasting the results based on learned system and salient features of given data which is indispensable for reliable operation.

5.3.2 Device Lifetime and Energy Challenge

In IoT research field, IoT-based intelligent devices and objects are related to the lifetime of the IoT device. This is a big challenge of computation of lifetime of devices. The lifetime mismatch of intelligent devices requires being deliberated in the full design and management of the different organization, configurations and installations involving IoT devices [9, 20, 21]. Moreover, management of energy and complete utilization of energy for intelligent devices are also the major challenging problem. It is especially for active IoT intelligent sensor devices [9]. Based on overall operations and computation, the energy harvesting can be a better solution for intelligent devices. The harvesting of energy can be done based on good predictive computation model or system [10].

5.3.3 Representation of Massive Data and Information Challenge

Representation of the massive amount of data and extraction of meaningful information is the important aspect of the current state-of-the-art-based IoT-based approaches. The IoT-based intelligent devices are significant to receive data from various sources (devices) [11, 12]. The received data are the source of richness and spatial distributed data, historical and sensor data. For example, taking a simple industry sensing and supervision as example, with 1000 intelligent devices or sensors connected to Internet and collecting sensing data by using these devices such as temperature, three-axis acceleration data with 10 K samples per s, and multimedia audio and video channel with 100 K samples per s, also considering that data is collected only 12% of daytime, 100 min per day, the total amount of received raw data for one year will be up 1000.4 PB/year [11, 12, 22]. This is a huge data. It is very difficult to store the data with different kinds of formats in the database. The predictive computation models are also unable to perform any computation and prediction on such amounts of raw data. It is needed to be processed and condensed and analysed in order to be functional at all so not data, but information behind it needs to be extracted for basic use [13, 23].

In the similar direction, the massive amount of data (volume and data heterogeneity) is also an important problem in the field of the IoT and data storage [13, 24, 25]. From the huge data, what data are required to perform the prediction and compute the performance of predictive modelling based computational devices? Which main factors and parameters contribute to the failure of predictive models? Based on observations, optimum maintenance schedule is also a significant problem in the IoT-based research field [26].

To solve these problems, new predictive model-based computation systems and efficient algorithms to visualize and represent data in the feature space [27, 28] are required. All this huge data and information also need special algorithms, mathematical models and meta-heuristics computational models, machine learning approaches regarding representation of data, extraction of features, handling and management in terms of data security, integrity and access mechanisms in the intelligent devices.

5.3.4 Incomplete Data Representation

Sparse, uncertain, complex and incomplete huge data are the discriminatory features for wide applications in big data prediction. Being data sparse, the number of feature points in given data is too less for making analysis and drawing reliable complete conclusions [9, 29, 30]. The feature space does not present complete patterns or data trends and distributions of data for different applications and predictive analysis. In the fields of pattern recognition and machine learning, models and data mining based representational systems, set of algorithms are also unable to represent the data in high-dimensional space, because sparse data representation significantly deteriorates the reliability of different predictive models and computation systems [29, 30]. General data mining and emerging machine learning, deep learning and computational intelligent methods can be applied for dimension reduction of the massive amount of data. After the dimensional reduction, the predictive model extracts features from reduced data sets. The extracted features are stored in the database. After that, feature selection algorithms are applied to choose the salient set of feature data from the stored information or data [30]. Data dimension reduction techniques perform the reduction of data and include additional set of feature points or sample data to improve the data scarcity using generic unsupervised machine learning methods and pattern recognition approaches [31]. In the similar direction, uncertain data is a special kind of sparse data sets. Each data field has longer deterministic value, however, it is subjected to few random/error distributions [32]. It is mainly linked to different domain-specific applications with inaccurate data readings and collections [9, 14, 33–35].

Learning capability is another major challenging problem in big data analytics. The existing data mining and the machine learning algorithms fail to perform the computation and prediction of outcomes using the massive amount of data. The learning models are bound to take the input data with respective memory space [29]. With the rising demands, the great proliferation of applications and wide uses and the future possibility of the implementation of the IoT in different fields, IoT-based predictive modelling and computation intelligent algorithms play significant roles in several disciplines such as medical sciences, e-health services, agriculture and weather forecasting with learning and prediction capability. The IoT-based models and frameworks have been proposed for monitoring of the users in real-time scenario and to cater better support if required [31].

IoT-based computation models and frameworks provide numerous benefits of easy to deploy mechanism over classical and standard networks with existing learning capability, enhanced information security during communication, quick access to records and proper utilizations of energy and harvesting of energy over existing Internet of Things based computing paradigm and frameworks [36].

5.4 IoT-Based Predictive Modelling

According to Gartner [1], the massive number of intelligent sensors is equipped with devices, coupled with the volume, velocity. Different kinds of structural design for IoT data present major challenges in the areas of data storage, security, data integrity, storage management, transferring of massive amount of data from source to destination server and the distributed data centre network in real-time scenario. Therefore, IoT can affect the computation and information sharing among intelligent objects on a different kind of computational paradigm. IoT-based predictive computation based models are required to design to solve these major challenges.

5.4.1 Predictive Model

The predictive modelling is defined as the process of applying a statistical model or data mining algorithm to data for the purpose of predicting the outcome of new or future observations or data sets. Moreover, in non-stochastic based predictive computation on data sets, the main objective is to predict the output Y for unknown observations $X1$, $X2$, $X3$, …, Xn. In given observations, it also includes temporal forecasting, where each observation is collected with respect to given time T. The inputs are used to forecast the result using future values at time $T + t$, where $t > 0$. In IoT-based predictive computation, the predictive model performs the predictions of future observations. The prediction of data includes point-wise or interval-wise predictions, prediction regions or prediction analysis based on the distribution of data, and ranking method for new observations. Predictive models select any method that produces predictions, regardless of its underlying approach. Moreover, predictive computing based models perform the process of information extraction using mathematical methods and the set of computational tools from existing data sets to identify patterns. The identification of patterns helps to predict future outcomes from given unknown data sets and trends in real-time scenarios.

5.4.2 *Descriptive Modelling*

Descriptive modelling is a type of modelling which is aimed at summarizing or representing the massive amount of the data structure in a compact representational manner. Unlike explanatory data computation modelling based frameworks, in descriptive modelling, the reliance is on an underlying causal theory which is absent or incorporated in a less formal way. Also, the focus is at the measurable level. Fitting a regression model for the given data can be descriptive if it is used for capturing the association between the dependent and independent variables rather than for causal inference or for prediction.

5.5 IoT-Based Predictive Techniques

In this section, IoT-based predictive methods are discussed. These methods use the coherent set of fundamental aspects of probability and statistics theory to solve major problems in big data.

5.5.1 *Probabilistic Learning Model and Statistical Analysis*

For uncertain data, the major challenge is that each data item is represented as sample distribution, not as a single value, so most existing data mining algorithms cannot be directly applied. Common solutions are to take the data distributions into consideration to estimate model parameters. For example, error-aware data mining [14] utilizes the mean and the variance values with respect to each single data item to build a Naive Bayes-based predictive model for classification. Comparable approaches have also been applied for decision trees or database queries. Incomplete data is defined as the missing of data field values for given samples or data [35]. The missing data values can be caused by different realities, such as the malfunction of a sensor node, or some systematic strategies to deliberately skip some data values (e.g. dropping some sensor node readings to save power for transmission) [37]. Most modern data mining approaches have in-built solutions to handle missing data values (such as ignoring data fields with missing values), data imputation (imputation of missing values to produce improved models) [21, 35, 37].

5.5.2 *Predictive Analytics and Data Mining Algorithms*

The predictive analytics and models apply the classification techniques for classifying the data. Classification is a machine learning technique to classify the data into different classes. The primary objective of a classification model is to predict a target variable (the unknown variable) using the similarity score based matching techniques. The classification method generates the outputs in binary forms (e.g. a loan decision for the bank customer) or particular (e.g. a client type) when a set of input variables are provided (e.g. credit score, customer name, income level, etc.). The process in data mining includes prediction of customer demands based on collected data in businesses. The classification model performs prediction of the massive data by applying the learning technique to find the generalized relationship between the predicted target variable with all the stored attributes of input data in the stored data set.

5.5.3 *Data Fusion Approach*

Data fusion is a computing paradigm to provide a technique to fuse multiple data received from different sensors or objects. In IoT, various sensors or intelligent devices are equipped with primary source model or framework; received data is unstructured and unorganized. The received data has to be fused. In the predictive computing models and maintenance cases, fusion approaches are applied to business data, telematics and original sensory data. In the fusion mechanisms, few data is obtained from the external sensors data sources, such as weather databases. These external databases are received by the Internet of Things based intelligent devices. After that, predictive computing machine takes the collected data and maps to feature space and generates similarity matching scores. Data fusion based predictive models and frameworks perform prediction and analysis of data using supervised and unsupervised machine learning techniques. Depending on the availability of explicit and implicit sensors links between diverse data sources, many machine learning methods and big data analytics techniques of connecting data are needed in IoT-based predictive computing models and framework to evaluate the experimental results [38, 39].

5.5.4 *Failure Prediction Using a Tree Ensemble Classifier*

For prediction of failures, records of sensor data are also validated using machine learning-based ensemble classifier techniques in the big data [40]. Tree ensemble classifier models are applied to learn sensory data to predict failures of the future system from past failures. The ensemble classification models use the gradient boosting classification approach to representing the data using regression trees. The regression tree is used to represent the classified data in the form of the tree. It builds an ensemble of classification trees one by one. After that, the predictions of the individual trees are added for the evaluation of final prediction score of given data. In the final regression tree, the consequent tree tries to build a model to generate the differences between the unknown data item and training data and the current ensemble prediction value by reconstructing the residual [40, 41]. The ensemble classification model trains a boosted regression tree framework. The framework consists of a series of trees that encodes characteristics of data records about failure. Based on the values of features of a given data record, the model is trained in such a way that each tree can decide which set of record groups a particular item belongs to. An appropriate weight is then assigned to each record, indicating evidence for or against the record belonging to a failing system. The model aggregates the evidence weights of all trees and outputs a probability of failure. Thus, the model reflects the likelihood that the given data record is an indication of a failing system [41, 42]. Figure 5.5 depicts a tree ensemble

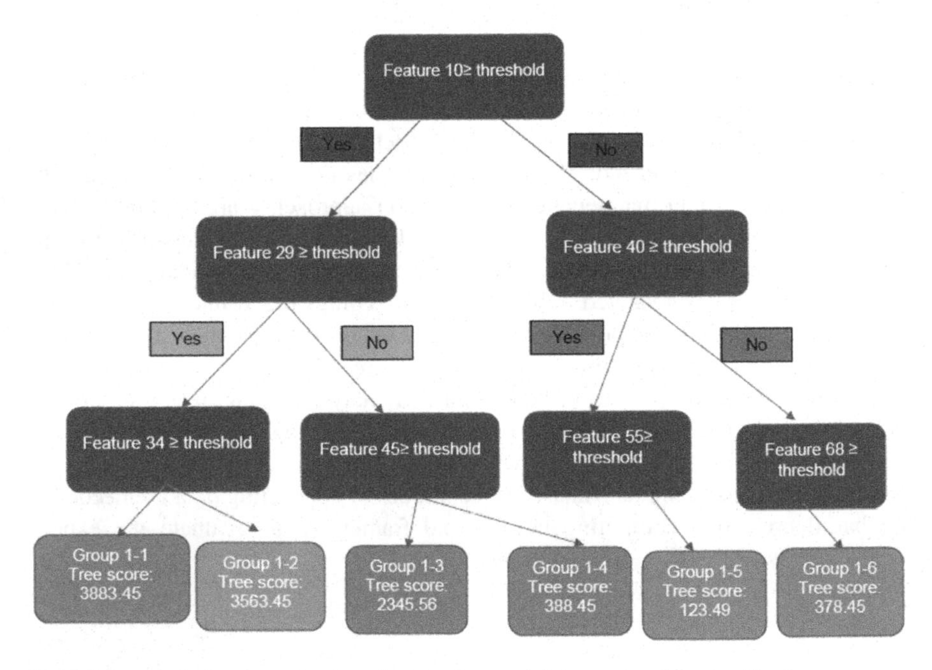

Fig. 5.5 Example for ensemble tree classification model for sensor data classification

classification model which is created by the regression-based classification algorithm. The regression tree based classification algorithm is a supervised learning method. It indicates for each record whether a record belongs to a regular or a failing system.

5.6 Summary

In this chapter, IoT-based predictive computing and modelling techniques are discussed. Due to advances in data storage, data computations, data capacity, machine learning, statistical modelling, etc., techniques based on predictive computing and modelling techniques have gained lot of popularity. A variety of applications based on these techniques are proliferating in the market. The predictive modelling and computation techniques are applicable to modelling credit card fraud detection, e-Health, e-Transportation, e-Logistics, e-horticulture, etc. These techniques collect the data in real time and provide the predictive information accordingly. However, IoT-based techniques face a lot of challenges which need to be handled carefully.

References

1. Stankovic JA (2014) Research directions for the internet of things. IEEE Internet Things J 1 (1):3–9
2. Moreno MV, Santa J, Zamora MA, Skarmeta AF (2014) A holistic IoT-based management platform for smart environments. In: Proceedings of communications 2014 international conference on (ICC), IEEE, pp 3823–3828
3. Atzori L, Iera A, Morabito G (2010) The internet of things: a survey. Comput Netw 54 (15):2787–2805
4. Botta A, De Donato W, Persico V, Pescapé A (2016) Integration of cloud computing and internet of things: a survey. Future Gener Comput Syst 56:684–700
5. Cai H, Xu B, Jiang L, Vasilakos AV (2017) IoT-based big data storage systems in cloud computing: perspectives and challenges. IEEE Internet Things J 4(1):75–87
6. Ning Z, Xia F, Hu X, Chen Z, Obaidat MS (2017) Social-oriented adaptive transmission in opportunistic internet of smartphones. IEEE Trans Ind Inform 13(2):810–820
7. Sharma SK, Wang X (2017) Live data analytics with collaborative edge and cloud processing in wireless iot networks. IEEE Access 5:4621–4635
8. Hui Y, Su Z, Guo S (2017) Utility based data computing scheme to provide sensing service in internet of things. IEEE Trans Emerg Top Comput. doi:10.1109/TETC.2017.2674023
9. Yang Y, Wu L, Yin G, Li L, Zhao H (2017) A survey on security and privacy issues in internet-of-things. IEEE Internet Things J. doi:10.1109/JIOT.2017.2694844
10. Johnston S, Apetroaie-Cristea M, Scott M, Cox S (2016) Applicability of commodity, low cost, single board computers for Internet of Things devices. In: world forum on internet of things (WF-IoT), pp 1–6
11. Williams PA, McCauley V (2016) Always connected: the security challenges of the healthcare Internet of Things. In: Proceedings of 3rd world forum on internet of things (WF-IoT), IEEE, pp 30–35

12. Alrawais A, Alhothaily A, Hu C, Cheng X (2017) Fog computing for the internet of things: security and privacy issues. IEEE Internet Comput 21(2):34–42
13. Cai H, Gu Y, Vasilakos A, Xu B, Zhou J (2016) Model-driven development patterns for mobile services in cloud of things. IEEE Trans Cloud Comput
14. Gubbi J, Buyya R, Marusic S, Palaniswami M (2013) Internet of things (IoT): a vision, architectural elements, and future directions. Future gener comput syst 29(7):1645–1660
15. Von Bischhoffshausen JK, Paatsch M, Reuter M, Satzger G, Fromm H, (2015) An information system for sales team assignments utilizing predictive and prescriptive analytics. In: Proceedings of 17th conference on business informatics (CBI), IEEE, vol 1, pp 68–76
16. Sampathirao AK, Sopasakis P, Bemporad A, Patrinos P (2017) GPU-accelerated stochastic predictive control of drinking water networks. IEEE Trans Control Syst Technol. doi:10.1109/TCST.2017.2677741
17. Parwez MS, Rawat D, Garuba M (2017) Big data analytics for user activity analysis and user anomaly detection in mobile wireless network. IEEE Trans Industr Inf. doi:10.1109/TII.2017.2650206
18. Chen Y, Chen H, Gorkhali A, Lu Y, Ma Y, Li L (2016) Big data analytics and big data science: a survey. J Manage Anal 3(1):1–42
19. Chong D, Shi H (2015) Big data analytics: a literature review. J Manage Anal 2(3):175–201
20. Bradley D, Russell D, Ferguson I, Isaacs J, MacLeod A, White R (2015) The internet of things-the future or the end of mechatronics. Mechatronics 27:57–74
21. Elmroth E, Leitner P, Schulte S, Venugopal S (2017) Connecting fog and cloud computing. IEEE Cloud Comput 4(2):22–25
22. Hou L, Zhao S, Xiong X, Zheng K, Chatzimisios P, Hossain MS, Xiang W (2016) Internet of things cloud: architecture and implementation. IEEE Commun Mag 54(12):32–39
23. Alam KM, Sopena A, El Saddik A (2015) Design and development of a cloud based cyber-physical architecture for the internet-of-things. In: Proceedings of ieee international symposium on multimedia (ISM), IEEE, pp 459–464
24. Xu B, Da Xu L, Cai H, Xie C, Hu J, Bu F (2014) Ubiquitous data accessing method in IoT-based information system for emergency medical services. IEEE Trans Industr Inf 10(2):1578–1586
25. Wang C, Bi Z, Da Xu L (2014) IoT and cloud computing in automation of assembly modeling systems. IEEE Trans Industr Inf 10(2):1426–1434
26. Al-Fuqaha A, Guizani M, Mohammadi M, Aledhari M, Ayyash M (2015) Internet of things: a survey on enabling technologies, protocols, and applications. IEEE Commun Surv Tutor 17(4):2347–2376
27. Yuan D, Jin J, Grundy J, Yang Y (2015) A framework for convergence of cloud services and Internet of things. In: Proceedings of 19th international conference on computer supported cooperative work in design (CSCWD), IEEE, pp 349–354
28. Wang C, Vo HT, Ni P (2015) An IoT application for fault diagnosis and prediction. In: Proceedings of international conference on data science and data intensive systems (DSDIS), IEEE, pp 726–731
29. Derguech W, Bruke E, Curry E (2014) An autonomic approach to real-time predictive analytics using open data and internet of things. In: proceedings of 11th international conference on ubiquitous intelligence and computing and autonomic and trusted computing. Proceedings of 14th international conference on scalable computing and communications and its associated workshops (UTC-ATC-ScalCom), IEEE, pp 204–211
30. Moreno MV, Terroso-Saenz F, Gonzalez A, Valdes-Vela M, Skarmeta AF, Zamora-Izquierdo MA, Chang V (2016) Applicability of big data techniques to smart cities deployments. IEEE Trans Industr Inf 13(2):800–809
31. Hossain MS, Muhammad G, Abdul W, Song B, Gupta BB (2017) Cloud-assisted secure video transmission and sharing framework for smart cities. Future Gener Comput Syst
32. Chen CP, Zhang CY (2014) Data-intensive applications, challenges, techniques and technologies: a survey on big data. Inf Sci 275:314–347

33. Raghupathi W, Raghupathi V (2014) Big data analytics in healthcare: promise and potential. Health Inf Sci Syst 2(1):3
34. Sagiroglu S, Sinanc D (2013) Big data: a review. In: Proceedings of collaboration technologies and systems (CTS), IEEE, pp 42–47
35. Ray PP (2014) Home Health Hub Internet of Things (H3IoT): an architectural framework for monitoring health of elderly people. In: Proceedings of international conference on science engineering and management research (ICSEMR), IEEE, pp 1–3
36. Gupta PK, Maharaj BT, Malekian R (2016) A novel and secure IoT based cloud centric architecture to perform predictive analysis of users activities in sustainable health centres. Multimedia Tools Appl. doi:10.1007/s11042-016-4050-6
37. Vermesan O, Friess P, Guillemin P, Gusmeroli S, Sundmaeker H, Bassi A, Jubert IS, Mazura M, Harrison M, Eisenhauer M, Doody P (2011) Internet of things strategic research roadmap. Internet Things-Global Technol Societal Trends 1:9–52
38. Singh D, Tripathi G, Jara AJ (2014) A survey of internet-of-things: future vision, architecture, challenges and services. In: Proceedings of world forum on internet of things (WF-IoT), IEEE pp 287–292
39. Yan Z, Zhang P, Vasilakos AV (2014) A survey on trust management for Internet of Things. J netw comput appl 42:120–134
40. Rokach L (2010) Ensemble-based classifiers. Artif Intell Rev 33(1):1–39
41. Abawajy JH, Kelarev A, Chowdhury M (2014) Large iterative multitier ensemble classifiers for security of big data. IEEE Trans Emerg Topics Comput 2(3):352–363
42. Bifet A (2013) Mining big data in real time. Informatica 37(1):15–20

Chapter 6
Cloud-Based Information Security

6.1 Introduction

Cloud computing facilitates on-demand service model and provides resources, information and software on sharable basis to their users. It supports heterogeneous connectivity of systems and can interact with each other at same time. The cloud computing provides various dynamically scalable resources as a service over the Internet. There are several economic benefits of using cloud computing as it reduces the overall expenditure and provides better performance and data storage capacity. However, there are still some potential challenges left to focus such as security, privacy and trust. The data that is being communicated between users, and cloud systems need to be secured from different threats and attackers. In [1], Shaikh and Haider have stated that one of the reasons why the cloud computing is not fully accepted by the users is the security. The users are always in fear of losing their data as well as privacy. They have identified various thrust area of cloud security and categorized them. Alassafi et al. [2] have emphasized on secure use of information technology (IT) to reduce risks and for further possible improvement of confidence and trust among the customers. They have stated that IT governance and information security governance (ISG) are two major factors for an organization to promote and use of cloud successfully. While implementing cloud security, security risks are associated with various infrastructure layers like application layer, virtualization layer, trust layer, authentication layer, access control layer, etc., the cloud computing can introduce the variety of risks and threats related to these layers. In [3], Tianfield has discussed about the various issues of security in cloud computing. They have analysed the cloud security requirements in terms of fundamental issues like trust, availability, audit, integrity and confidentiality. As the security is a major issue, it should be applied at different levels to ensure right implementation of cloud computing such as: security of host server, security of data storage, network security and security of application. In [4], Gugnani et al. have focused on cloud-based web services and proposed an approach for selective encryption of

P.K. Gupta et al., *Predictive Computing and Information Security*,
DOI 10.1007/978-981-10-5107-4_6

XML elements. They have used deoxyribonucleic acid (DNA) encryption technique for selective XML elements.

The cloud architecture is classified into four different layers [5] as shown in Fig. 6.1:

(a) Physical layer—The hardware level resources like network resources, computing resources and storage resources are all contained in this layer.
(b) Resource layer—All the resources that have been virtualized lie in this layer. These resources can be used by upper layers and end users further.
(c) Service layer—It supports various services, and software tool, middleware and provides development and deployment of platform.
(d) Application layer—It consists of all the executable applications in the clouds.

The cloud computing models are generally classified into three service models and four deployment models. The service models, as shown in Fig. 6.2 which can be categorized into further three categories: (i) Software as a Service (SaaS), (ii) Infrastructure as a Service (IaaS) and (iii) Platform as a Service (PaaS). On the other hand, the deployment models typically consist of (i) private cloud, (ii) public cloud, (iii) hybrid cloud and (iv) community cloud.

6.2 Related Work

Though the cloud computing architecture and its models are widely adopted by the industries, it still has certain drawbacks. The foremost issue in cloud computing is of security and privacy related to the data of the users. The cloud computing model leaves the clients vulnerable to different types of attacks and threats. Due to this, the client may suffer from a heavy loss of any confidential data or may lose any confidential information. An attacker may eavesdrop the conversation between two clients on cloud or between client and cloud. The users who move their data onto

Fig. 6.1 The cloud architecture [5]

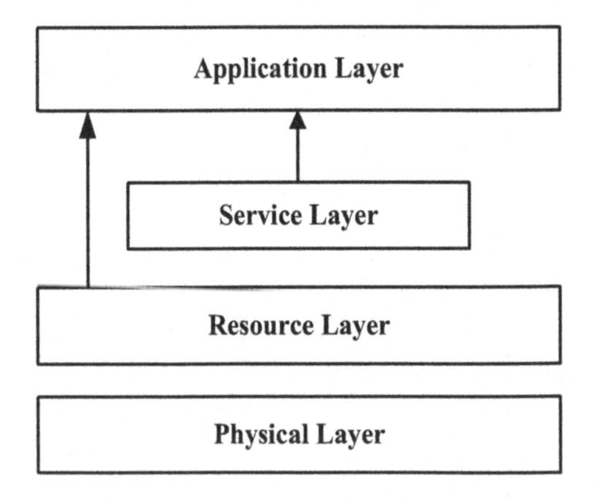

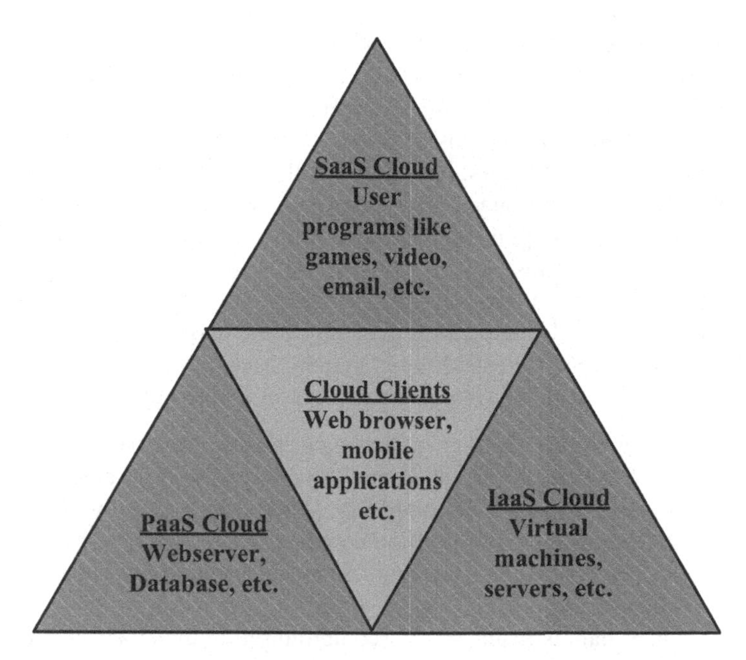

Fig. 6.2 The cloud computing service models

the cloud are unaware of the integrity of their data. They do not know as to where their data is getting stored. Due to these reasons, some users are still adamant of making use of this technology. So, there is a great need to protect the clients from such type of attacks.

The cloud security refers to a broad variety of technologies, policies, mechanisms, frameworks and controls deployed to protect data, applications and the associated infrastructure of cloud computing model. The potential areas that require focus to embed security features are:

- To safeguard all end cloud—user activities, actions regardless of device,
- To protect cloud, database and data centres, and
- To facilitate superior cyber security against various attacks.

The cloud security mechanism must ensure the implementation of correct defensive implementations. The well-organized cloud security mechanism should recognize and address the various security-related issues. Implementation of effective security mechanism and controls defend the system from flaws and decrease the possibility of an attack. These security controls can be categorized under the following category [1]:

- **Deterrent controls**—are anticipated to decrease attacks on a cloud. These deterrent controls typically diminish the effect of threat by notifying the attackers that there will be poor consequences for them if they continue further or move forward in that particular direction.

- **Preventive controls**—toughen the system against threats and attacks, and vulnerabilities. However, the well-built authentication of cloud customer stops unauthorized cloud access by the customers.
- **Detective controls**—are proposed to sense and respond to any accidents that occur. When an attack occurs, this control will sign the anticipatory or remedial controls to tackle the issue. The system and network sanctuary monitoring, intrusion detection and prevention arrangements, are typically engaged to sense attacks on cloud systems.
- **Corrective controls**—these controls decrease the result of a threat, usually by restricting the harm. These controls generally come into effect during or after an attack has occurred. Re-establishing system backup so as to reconstruct a cooperated system can be seen as a paradigm of a corrective control.

Since accessing of information in heterogeneous environment at a time, the integrity and privacy can be breached and always at risk. The cloud computing provides distribution of data over computers. When data is sent by the user to be processed in the cloud; the control of the data is given to a remote party that may not address security concerns of the user. As a user has no physical access to the data, he/she is unaware about the location of his/her data and is not sure whether the integrity of his/her data is maintained or compromised in the cloud. It is important to ensure that the information being processed on cloud is secure and no tampering of information is done when previously unknown parties may be present [6].

6.2.1 Security Issues

Security is termed as the prevention of any unauthorized access, unauthorized deletion or amendment of the information. ISO 27001 defines the security as: '*Preservation of confidentiality, integrity and availability of information; in addition, other properties such as authenticity, accountability, non-repudiation, and reliability can also be involved*'. Cloud computing may introduce many risks to cloud service and cloud deployment model. In [7], Aluvalu and Mundane have presented various access control techniques and models and stated that privacy, trust and access control are important factor to maintain the security in cloud. Rashdi et al. [8] have defined the term cloud computing security as, '*The set of control-based technologies and policies designed to follow to regulatory compliance rules and protect information, data applications and infrastructure associated with cloud computing use*'. The main dimensions of security that should be kept in mind for providing user satisfaction are confidentiality, integrity and availability, (CIA) [9]. CIA provides the basis for implementing security principles, as shown in Fig. 6.3, to known set of threats and can be known as follows:

- Confidentiality—stands to keep the user's data secret. According to Xiao and Xiao [10], confidentiality is one of the major issues in cloud because the information that is outsourced by users on cloud servers is managed and

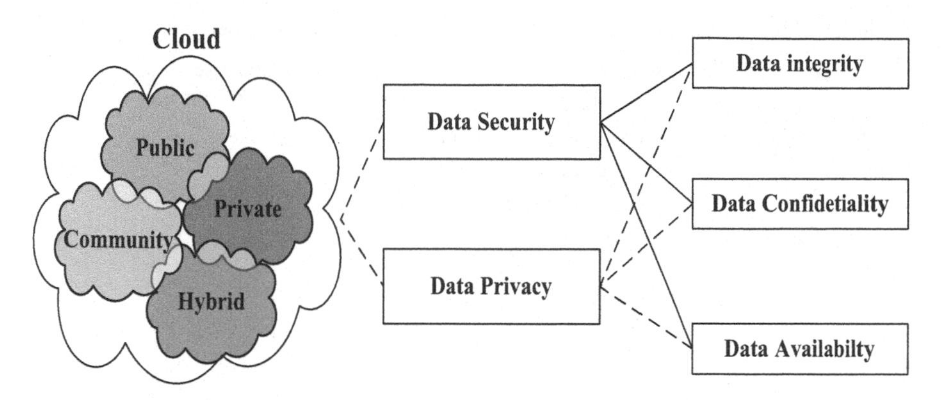

Fig. 6.3 Organizing security and privacy for various cloud deployment models

controlled by untrustworthy cloud providers. In [3], Tianfield also mentioned that threat to data increases because of increased number of applications, parties and devices which leads to increase in number of point of access. One way to achieve confidentiality is to encrypt the information sent by user before placing it in cloud.

- Integrity—stands to preserve the integrity of the information. It should be checked that the information is not lost or modified by unauthorized user. One technique to maintain the integrity is usage of digital signature. In [3], the authors have stated that by using service level agreement (SLA), information is protected while it is on cloud, preventing intrusion or attack on data and responding swiftly to attacks such that damage is limited.
- Availability—The users can access the resources, information from any place and at any time. Denial of service attacks, equipment outages and natural disasters are all threats to availability of information in clouds. According to Zissis and Lekkas [12], the availability should not only be in terms of information, software but also hardware being available to authorized users upon demand.

The risks associated to stored information in clouds can vary according to cloud service models deployed by the organization. Some risks can affect all the cloud service models including SaaS, PaaS and IaaS, whereas some are limited to one or two models only. Table 6.1 represents some of the major risks of cloud computing associated to CIA security principles.

6.2.2 Privacy Issues

The cloud computing uses different ways to manage the information and user-related personal data. The privacy refers to the right to self-determination, i.e. the ability of individual or group to seclude them from access of information and

Table 6.1 Various risks associated to CIA security principles and cloud service models

CIA security principles	Associated risks	Cloud service models		
		SaaS	PaaS	IaaS
Confidentiality	Users from organization	•	•	•
	Attacker(s)	•	•	
	Data leakage	•	•	
Integrity	User access	•	•	•
	Data segregation	•	•	
	Quality of data	•	•	
Availability	Change of policies or management	•	•	•
	Denial of service	•	•	•
	Physical disruption			•
	Lack of recovery methods	•	•	•

then selectively reveal them. Privacy issues are becoming more important while using Internet-based transactions. Lack of effective security mechanism and loss of control could result in serious threat to data integrity, confidentiality and privacy principles [13]. In terms of organization, personally identifiable information is managed by providing privacy which involves the application of processes, standards, laws and mechanisms. Xiao and Xiao [10] have considered the emerging cloud platform and used an attribute-driven methodology to design security and privacy paradigm. They have used attributes like confidentiality, availability, integrity, accountability and privacy-preservability. In [14], Chen and Zhao have analysed the data security and privacy-related issues using various algorithms. They have stated that because of these issues, many large organizations still do not share their data on cloud. In [15], Sun et al. have identified the various elements related to privacy and categorized them in three groups known as (i) *When*—a user is more cautious about the use of information which is either being accessed or will be accessed, (ii) *How*—user must ensure the way for accessing this information manually or automatically, (iii) *Extent*—user can define the several points as an ambiguous region and can only be used by group of users those who have the precise access to that region. Privacy issues vary according to different cloud scenario. In [16], Pearson has mentioned that existing cloud services impose a lot of challenges to privacy of data while handling sensitive data, and data leakage related issues and suggested that this type of data cannot be stored in public clouds of various cloud service providers in an unencrypted form. Pearson [16] and Guilloteau et al. [13] have identified the key potential privacy issues as follows:

- Lack of user control—As the user stores their data over public cloud, then cloud service providers become responsible for further handling and managing of data. Users have limited control over the stored data in public clouds. There is always an issue of transparency in between the cloud user and service providers as the stored data over clouds can be analysed by security agencies or law enforcement agencies. A trust is required between both of the parties while transferring data rights and providers must ensure the data integrity throughout its lifecycle.

- Lack of training and expertise—Developers are focusing more on easy to design and deployment models of clouds rather the handling of privacy concerns to their models. They need to be trained with existing information security laws and practices to maintain the privacy of data. By ensuring privacy of users data, one can build the trust over the period of time.
- Unauthorized use of data by third party—There is always a risk associated with the stored data in the clouds that it can be used by the third party for their own purpose. Nowadays, it is getting more common as users may get annoying advertisements while accessing or using of stored information. Currently, there is no such arrangement to stop this unauthorized use of data by third party.
- Achieving regulatory compliance—Global use of cloud computing makes it complex as user never knows about the exact location of data that is being stored in clouds. These cloud servers can be in the same country or may be located globally. Things become complex if cloud is located in different country as many legislations in place around the world. It is always difficult to ensure compliance with all the legislations. The cloud computing may exacerbate the transborder data flow issue that may restrict the flow of information.

6.2.3 Trust Issues

Trust is a measureable belief that is used to make trustworthy decisions based on experience and it is a major issue with cloud computing irrespective of the cloud model being deployed. The security and privacy challenges discussed above are also relevant to the general requirement upon cloud suppliers to provide trustworthy services. Trust relationships are very much at the centre of certain security and privacy solution. To build trust, one need to ensure security and privacy of data is intact and foolproof. This situation is also depicted in the Fig. 6.4, where trust lies in between security and privacy. Trust has several different attributes like reliability, confidence, dependability, honesty, etc., to obtain various cloud services. Sun et al. [9] and Pearson [16] have identified various issues of trust in cloud computing:

- The attributes of cloud computing environment are unique, so the definition and evaluation of trust become difficult.
- Based on the degree of trust, how to provide different security level of services.
- The trust relationship in cloud computing is temporary and dynamic, so handling of malicious information is a tough task.
- Lack of consumer trust is another major reason for avoiding the cloud adoption. Various critical challenges like vendor's lock-in, cloud availability, cloud performance, cloud data security, etc., need to be addressed to encourage cloud adoption. In clouds, customers have limited control of resources that is why they cannot protect their data against unauthorized access or its misuse.

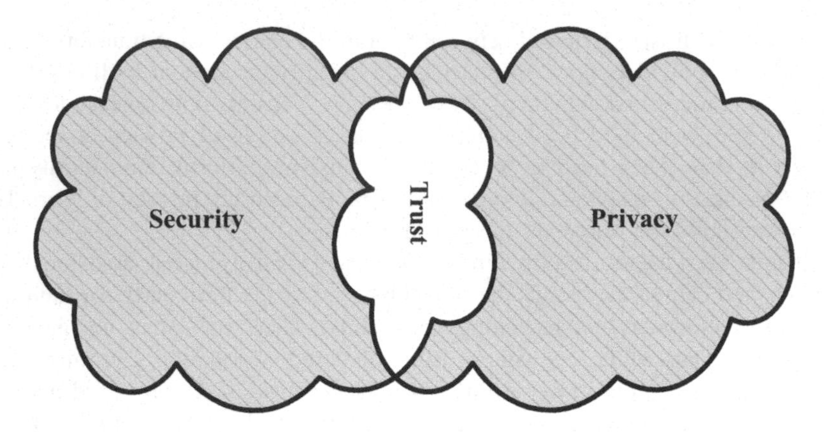

Fig. 6.4 Obtaining trust from security and privacy of data

- Weak trust relationships and lack of transparency between cloud user and service provider may lead to failure of cloud service delivery chain.
- Lack of knowledge and consensus for using standardized trust models and approaches for cloud environment [17].

6.2.4 Issues with Cloud Service Models

Apart from the issues as stated in the preceding sections, there are many other issues of cloud computing pertaining to its different service models known as IaaS, PaaS and SaaS. In [18], Rana et al. have proposed a combined and improved IaaS and PaaS architecture to remove their drawbacks.

- **IaaS issues**—It provides only basic level of security like load balancing, firewall, etc. The applications moving to the cloud require high degree of security. One company may be hosting many other companies' workloads and data in a shared environment. In such cases, it may expose all parties to a higher risk of security- or privacy-related incidents [19]. In [20], Joshi et al. have listed the component wise security issues in IaaS like SLA issue for monitoring of quality of services to cloud users and enforces trust between cloud service provider and user, and cloud software related issues which coins the cloud component together to make them act as a single component so that attackers cannot easily target the SOAP envelop or XML signature. The different issues in IaaS are summarized as follows in Table 6.2 [21]:
- **PaaS issues**—The hacker can use the advantages of PaaS to influence the PaaS cloud infrastructure for malware command and control. One major challenge is the interoperability of PaaS as most of the applications, APIs, database are vendor specific. The various issues in PaaS are summarized in Table 6.3 [21].

Table 6.2 Issues in IaaS

Area	Security	Privacy	Trust
DOS (denial of service)	Misconfiguration, vulnerabilities in system or OS	Access control compromised	Service not available
Robustness of virtual machine-level Isolation	Vulnerabilities in hypervisor	Internal network probing may occur	Compromised virtual machines/hypervisors permit the loss of trust
EDOS (economic denial of service)	Authentication, authorization, and accounting (AAA) vulnerabilities	User provisioning, de-provisioning vulnerabilities	Access control compromised

Table 6.3 Issues in PaaS

Area	Security	Privacy	Trust
Technical immaturity	Compliance challenges	Storage of data in multiple jurisdiction and lack of transparency about this	Lack of information on jurisdictions
Lack of portability	Non availability of common authentication interface	'Data hostage' clause in supplier outsourcing contracts	Liquidate damage for lost business
Protecting API keys	Bad key management procedures	Service information leakage	Lack of sensitivity

- **SaaS issues**—The main focus is not only on application's portability but also on migration of data and enhancement of security functionalities. For development and deployment of SaaS application, the order of following security elements must be ensured. The following should be kept in mind:
  - *Data security*—In SaaS model, the data is stored at the vendor's end. The SaaS vendor should acquire additional security checks so as to ensure security of data and prevent data breach through unauthorized users. The strong encryption techniques should be involved for data security. Due to loophole in data security model, malicious users can gain access to the data.
  - *Network security*—The sensitive data that is obtained from the users/organizations are stored at SaaS vendor end. All the data over the network must be protected to prevent leakage of data. This is achieved by using strong encryption techniques to manage network traffic like SSL and TLS.
  - *Data locality*—The applications provided by SaaS are used by consumers and then data processing is done. The consumers are unaware of the fact as to where their data is getting stored. Example: In many European countries, certain type of data cannot leave the country because of the information being sensitive. The SaaS model should provide reliability to the customer in terms of location of data
  - *Data integrity*—It is easier to achieve data integrity in a single system with single database by making use of database constraints and transactions. Transactions follow ACID properties. The problem of data integrity gets magnified in case of cloud computing. The SaaS vendors unveil their web service APIs without any support for transactions. There are different levels of availability and SLA in each SaaS application which makes it difficult to manage the transactions and provide data integrity
  - *Data segregation*—Due to multi-tenancy feature in cloud computing, multiple users can store their data on cloud. The data of various users will reside at same location. Intrusion in user's data by another user becomes easy in such environment. Intrusion can be done by hacking or by injecting client's code into SaaS system. Therefore, SaaS model should ensure boundary for each user's data not only at physical level but also at application level

Table 6.4 Issues in SaaS

Area	Security	Privacy	Trust
Unauthorized access	Data integrity and confidentiality loss	Compromised communications secrecy	Loss of trust in service
Physical risks	Physically destroyed data	–	–
Browser-based risks	Loss of data integrity, and confidentiality	Loss of user secret credentials	Loss of confidence upon channel
Network dependence	Loss of availability	–	Trust on service reduces

- *Data access*—Various security policies are provided by the organization to the users when accessing the data. Based on these policies, each employee can access limited information. Cloud must stick to these security policies in order to avoid intrusion of data.
- *Authentication and authorization*—The authentication is assurance that the communicating entity is the one that it claims to be that is 'who are you?' and authorization is a kind of access control 'what you can do'. This means that no unknown or harmful entity should able to pretend that he or she is the authenticated one and even authenticated persons should have a limited access to the data.

The various issues in SaaS are summarized in Table 6.4 [22].

6.2.5 Threats in Cloud Computing

The cloud security alliance has presented a primary draft for threats relevant to the security architecture of cloud services. In this section, we have given few potential threats related to the cloud [23, 24].

(a) **Malicious insiders**—Majority of the companies conceal their strategies about the height of access to their staff. Though, via superior level of access, a member of staff can grow access to top secret data and services. As there is deficiency in transparency of cloud provider's policies, processes and procedures, some insiders can frequently have the privilege to access the client's data. Malicious insiders (employee) actions are often evaded by a firewall or infringement discovery system considering it to be an authorized action. Though, a trusted member of staff may also convert into an opponent. In these kinds of scenarios, insiders can source a significant effect on cloud services. Let us take an instance —here malicious insiders can access top secret data and put on control over the cloud services without any jeopardy of revealing his identity. These kinds of threats may be applicable to any cloud service SaaS, PaaS and IaaS. So as to avoid these kinds of risks, there is the need of more transparency in security and management process together with compliance reporting and breach notification. This is amplified in the cloud via the meeting of IT services and client over a single management domain which is united within a general transparency deficiency into the service supplier process and procedure [25].

(b) **Shared technology issues/multi-tenancy nature**—Virtualization in multi-tenant architecture is used to provide shared on-demand services. Different users who have access to the virtual machine may use the same shared application. Though, as mentioned above, via some attacks and threats, some malicious entities can gain access and control of the lawful users' virtual machine. In multi-tenant architecture through shared resources, IaaS services are delivered which sometimes are not designed to give sufficiently strong isolation. Giving permission to one tenant to interfere in the other can cause serious affect on the cloud architecture which can affect its regular operations. Generally, these types of threats have an effect on IaaS. Transparency in SLA for patching, well-formed authentication system and access control mechanisms to administrative tasks are some of the solutions to resolve this issue.

(c) **Insecure Interfaces and APIs**—Cloud providers offer their services by using the various types of APIs, available for different cloud service models like SaaS, PaaS and IaaS. These weak set of APIs and interfaces can result in many security-related issues in cloud which includes unauthorized access of security key and data, insufficient input data validation, weak credentials that lead to loss of data integrity and CIA [26, 27].

(d) **Data threats**—In [27], Kazim and Zhu have stated that in cloud computing, data is the valuable resource for any organization, and security of the data is the biggest challenge for cloud service providers. Major data security threats that may arise over the period of time are data breaches, data loss, unauthorized access and integrity violations.

- **Data breaches**—are also commonly known as data leakage of stored information in cloud via unauthorized access. The major reasons of data breaches are poor application designing, flaws in operational and infrastructure setup, insufficiency of authentication, authorization and audit controls [27].
- **Data loss**—Information can be exchanged in many ways. This can incorporate data insertion, data compromise, deletion or modification. Data loss occurs because of many reasons like malicious attackers, data deletion, data corruption, etc. [27]. As the cloud is shared and dynamic in nature, so these threats could prove to be of a foremost concern leading to data theft.

(e) **Service hijacking**—is a very serious threat. In this, the hijacker may forward the cloud customer to an illegal website. For attackers, service instances and user accounts can serve as a new base for attack. This threat can have great impact on IaaS, PaaS and SaaS. There are some of the solutions to resolve this threat which include safety policies, strong authentication and activity monitoring.

(f) **Identity theft**—The identity theft is a kind of scam in which one acts as if to be someone else so as to access resources or obtain credit and other benefits. The casualty (of identity theft) can undergo undesirable consequences and losses and held responsible for the criminal's activity. Some of the security perils are phishing attacks, weak password, recovery workflows, key loggers, etc. This threat can have significant impact on SaaS, PaaS and IaaS and some of the solution includes the use of powerful encryption, authentication and authorization mechanisms.

(g) **Denial of Service (DOS)/Distributed Denial of Service (DDOS)**—DOS attacks prevent users from accessing of cloud services, network and other services whereas in DDOS, attackers use the multiple network sources to send a large number of requests to the cloud for consuming its resources [27, 28].

(h) **Online Cyber theft**—In [29], Chou has stated that as the cloud-based services are becoming popular among the customers and organizations for storage of information, stored sensitive data is becoming an attractive target to online cyber theft. Online cyber thieves can make the use of stolen password and user IDs or can take the advantage of computing power offered by cloud service providers to launch attacks or to access user accounts.

6.2.6 Attacks on Cloud

There is number of security risks and issues as shown in Fig. 6.5, associated with cloud computing but they are grouped in two categories: (i) security issues faced by cloud provider and (ii) security issues faced by their customers. Various types of attacks in cloud are listed as follows:

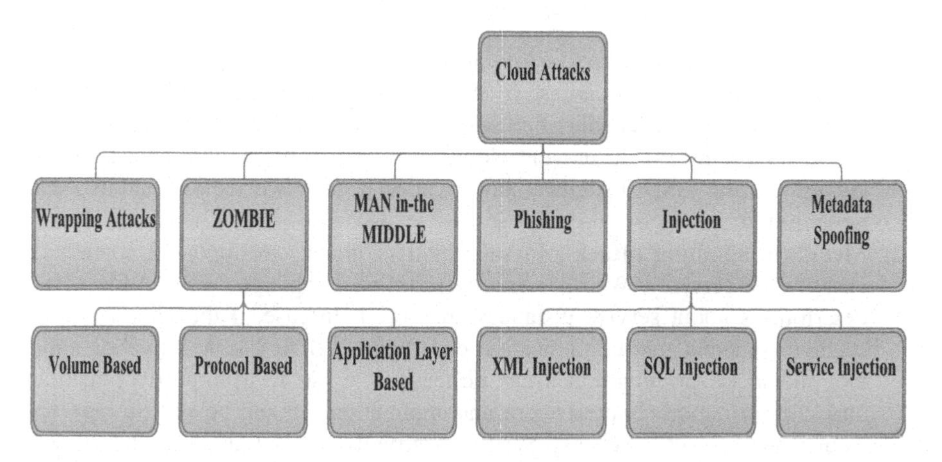

Fig. 6.5 Various types of cloud attacks

(a) **Zombie Attack**—An attacker can overflow the huge number of requests via *zombies*. This kind of attack disrupts the normal performance of cloud, disturbing the availability of cloud services. This cause cloud to be overloaded to serve a huge number of requests and therefore all its resources get exhausted, which can further originate DoS (Denial of Service). In DoS, in cloud due to the presence of invader's overflow of requests, cloud becomes unavailable to serve legitimate user's requests. Nevertheless, strong authentication and authorization and IDS/IPS can offer defense in opposition to such type of attack.

(b) **Injection attack**—Injection attacks are such attacks where intentionally malicious data is included as the input to disrupt the normal functioning of the cloud. A few of the injection attacks are:

- *XPath Injection:* When user and the cloud communicate, they communicate through XML files. XPath is a kind of query language for XML document as for relational databases we have SQL. Dissimilarity among SQL and XPath is that XPath is implementation independent [30]. XPath injection can occur by querying the XML database or when a service is invoked.
- *SQL injection:* In such kind of attacks, user injects malicious SQL statements into an entry field for execution. SQL injection is generally recognized as an attack vector for websites but can be used to attack any type of SQL database.
- *Service Injection:* An opponent attempts to insert a suspicious service or new illegal virtual machine into the cloud system and could supply malicious service to users. Cloud malware deforms the cloud services by changing (or blocking) cloud functionalities. Let us take an example in which an opponent makes its own malicious services like SaaS, PaaS or IaaS and attach it to the cloud system. In case, an opponent becomes successful to perform this, then legitimate requests are readdressed to the malicious services automatically. To protect against such an attack, there is a need to implement the service modules.

(c) **Man-in-the-Middle attack**—An attacker is capable of accessing the data exchange between two parties, if SSL is not configured properly. In case of cloud, an attacker is capable of accessing the data communication between data centres. To reduce or to prevent cloud from man-in-the-middle attack, proper configuration of SSL and data communication tests between cloud and its client are required.

(d) **Metadata spoofing attack**—In such kind of attack, an opponent amends or changes the service's Web Services Description Language (WSDL) file where descriptions about service instances are stored. In case, if the opponent succeeds to suspend service invocation code from WSDL file at delivering time, then metadata spoofing attack can be feasible. So as to triumph over such an attack, information about services and applications should be kept in ciphered form. Currently, WS-Security Service is broadly used in cloud to endow with security for the system [31].

(e) **Phishing attack**—is famous for manipulating a hyperlink and sends false link to obtain confidential information from the user. In cloud, sometimes it might be possible that an adversary uses the cloud service to host a phishing attack site to hack accounts and services of other cloud users.
(f) **Wrapping attack**—In [29], Chou has mentioned that web service security mechanism is used to ensure the confidentiality and integrity of SOAP messages communicated between user and cloud service providers. Wrapping attacks use XML signature wrapping to exploit a weakness when web servers validate signed requests [32]. These attacks generally take place during the translation of SOAP message between users and web servers. Since cloud users normally request services from cloud computing service providers through a web browser, wrapping attacks can cause damage to cloud systems as well. In [33], a group of researchers have demonstrated the use of XML signature wrapping technique for hijacking of an account that exploited vulnerability in the Amazon web service.

6.2.7 Cloud-Based Information Security Models

To ensure cloud security at various levels from information retrieval to its storage, security models are implemented. These security models must ensure the support for cloud security issues related to security issues, privacy issues, and trust issues as discussed in previous section. These security models also build the user's confidence by deploying various techniques for handling the various cloud-related threats. In [34], Mushtaq et al. have proposed the quad-layered framework for data security, data privacy, data breaches and process associated aspects. This proposed layered framework prevents the confidential information and try to build user's trust on cloud computing. In [35], Kritikos and Massonet have proposed a security meta-model for clouds. This model captures the high-level and low-level security requirements and capabilities to derive application deployment. Nafi et al. [36] have proposed a security model for cloud and implemented various security algorithms like RSA, AES and MD5 for secure communication of information between users and servers. To resolve the problem of privacy in the clouds, Metri and Sarote [37] have introduced a threat model known as STRIDE which helps in analysing a problem, designing appropriate strategies and evaluating the solutions. They have listed the following steps to ensure the privacy:

(a) Identify—attacks, and threats
(b) Prioritize threats—according to the impact using STRIDE model.

- Spoofing identity—means an attacker poses as another user or a machine poses as a valid machine.
- Tampering with data—means to maliciously modify the data.
- Repudiation.
- Information disclosure—means to expose the information to the unauthorized users.

- Denial of Service (DoS)—means to deny any service to valid users. Example: web browser made temporarily unavailable.
- Elevation of privilege—means the privileged access is gained by unprivileged users to destroy entire system.

(c) Select appropriate strategies for threats

In [38], Mathew has presented a model to help the cloud users and cloud providers to ensure the safety of data. He has used the secured virtual private network (VPN) for accessing clients and providers. In [39], a privacy protection framework was proposed by Gajanayake et al. based on information accountability (IA) components. They have used IA agent for accessing of information by the users. In case of any misuse of information, IA agent implements various methods to stop this. In [40], Chidambaram et al. have proposed a secure storage system to handle sensitive data in clouds. They have introduced a user authentication mechanism to prevent unauthorized access to data. RSA algorithm has been used for encryption of files in cloud and MD5 has been used to generate the digital fingerprint. Hamlen et al. [41] have proposed a framework consisting of layers for cloud security and focused on two of the layers known as the storage layer and the data layer. A bottom-up approach to security was also proposed, where work is done on small problems in the cloud to solve the larger problem of cloud security. First, the various methods have been discussed for securing of documents. Second, how security can be enhanced by using secure coprocessors is discussed. Juels et al. [42] described a formal 'Proof of Retrievability' (POR) model as shown in Fig. 7.6 for ensuring the remote data integrity. This model utilizes the error correction code approach to verify the validity of data. In POR protocol, the verifier stores only a single key for each file. This scheme requires that only a small portion of file F is accessed by the prover in course of a POR. POR encrypts file F and randomly valued check blocks called *sentinels* are embedded. The prover is challenged by the verifier by specifying position of collection of sentinels and asking the prover to return associated sentinel values. If any modifications are made on file F by the prover, then large number of sentinels is likely to be compressed. To protect corruption of file F by the prover, error correcting codes are employed.

In Fig. 6.6, by using encoding algorithm, raw file F is transformed into file F' and is stored with the prover/archive. A key K is produced by using the key generation algorithm and it is stored by the verifier. The key K is used in encoding mechanism. A challenge–response protocol is performed by the verifier with the prover to check that file F can be retrieved by the verifier. The drawback of the approach is that there is computational overhead. In [21], Mohta et al. have discussed about the implementation of TPA for verifying the cloud service provider (CSP). A protocol was proposed that is used for supporting various operations which also maintains privacy and integrity. The architecture is shown in Fig. 6.7

In [43], Chalse et al. provide a detailed analysis of the cloud security problem. The different problems in a cloud computing system and their effect upon the different cloud users are also analysed. This architecture is shown in Fig. 6.8 and consider the following:

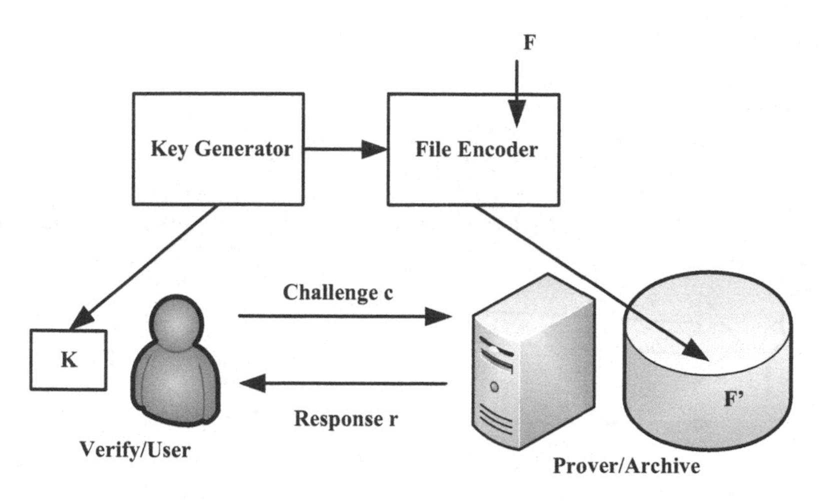

Fig. 6.6 Schematic of a POR system

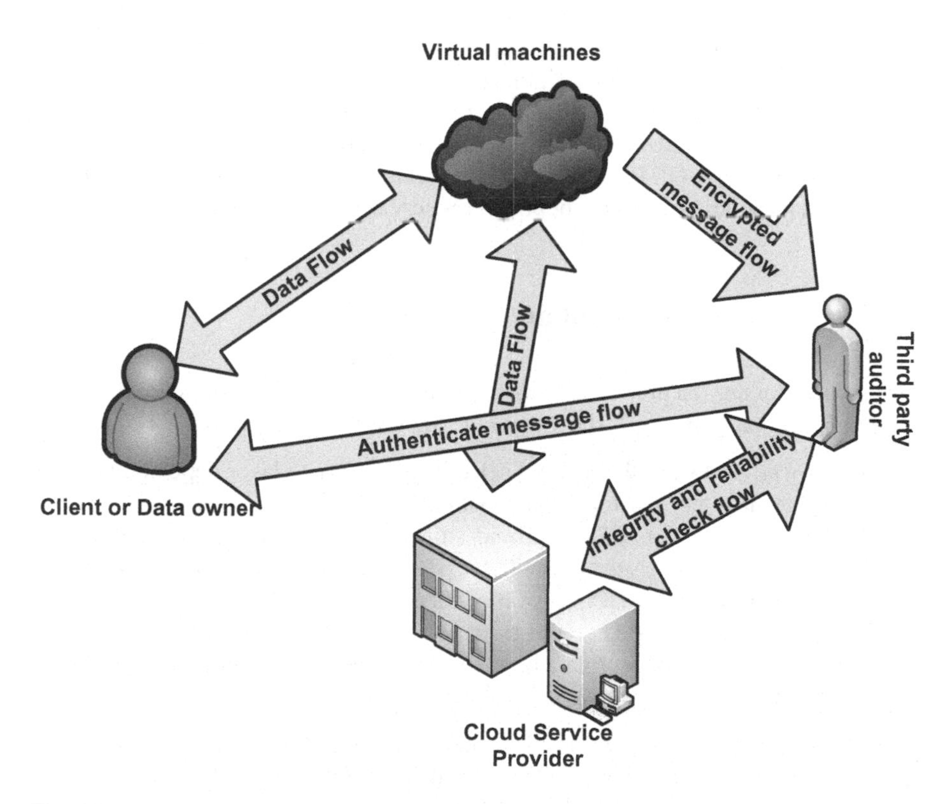

Fig. 6.7 Architecture for client, third party auditor and cloud service provider

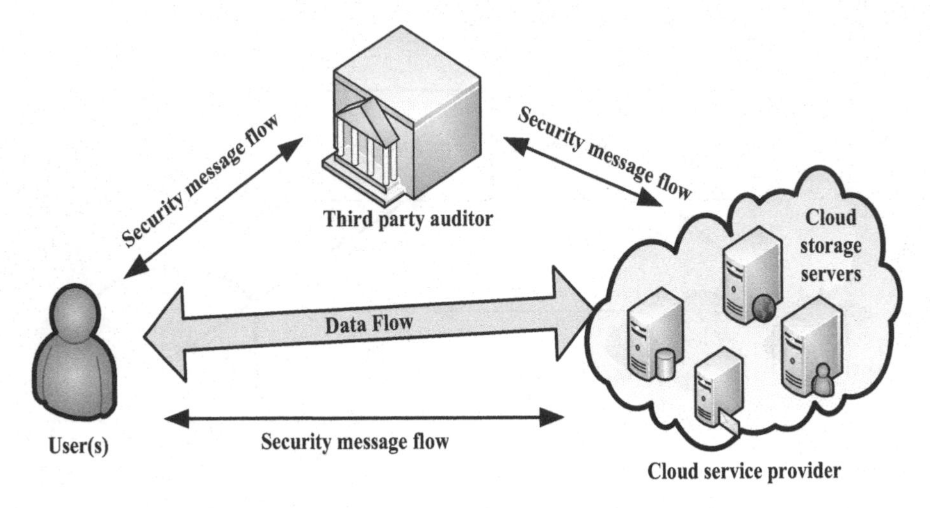

Fig. 6.8 Cloud data storage architecture

- A client who wants to store large amount of data in multiple clouds and has the permissions to access and manipulate stored data.
- Cloud service providers (CSP) support various services related to data storage and its usage and have sufficient resources available for computation.
- Trusted third party (TTP) ensures various parameters related to storage and use of data and provides appropriate public query services available to these parameters.

6.3 Framework to Maintain Data Integrity

This section defines the proposed framework to provide data integrity in multi-cloud system. Proposed framework shown in Fig. 6.9 has three main roles [11]:

- *Users*—who will store the data by selecting appropriate layer depending on the level of security needed for the data stored on cloud.
- *Cloud service provider (CSP)*—provides the storage of data service with flexible resources to keep the user data. The CSP manages cloud server (CS) which informs the user about the intrusion of data on cloud.
- *Third party auditor (TPA)*—verifies the cloud server and checks whether there is any manipulation of user data by the cloud server. It then sends a report to the user stating that the cloud server (CS) was trusted or not.

There are many cloud service providers and each of them provides different storage plan along with different QoS parameters so it becomes a tough task for users to keep moving their data from one cloud to another based on QoS and cost optimization [14]. In the proposed model, concept of multi-cloud is used to provide best cost optimization for various requirements of user.

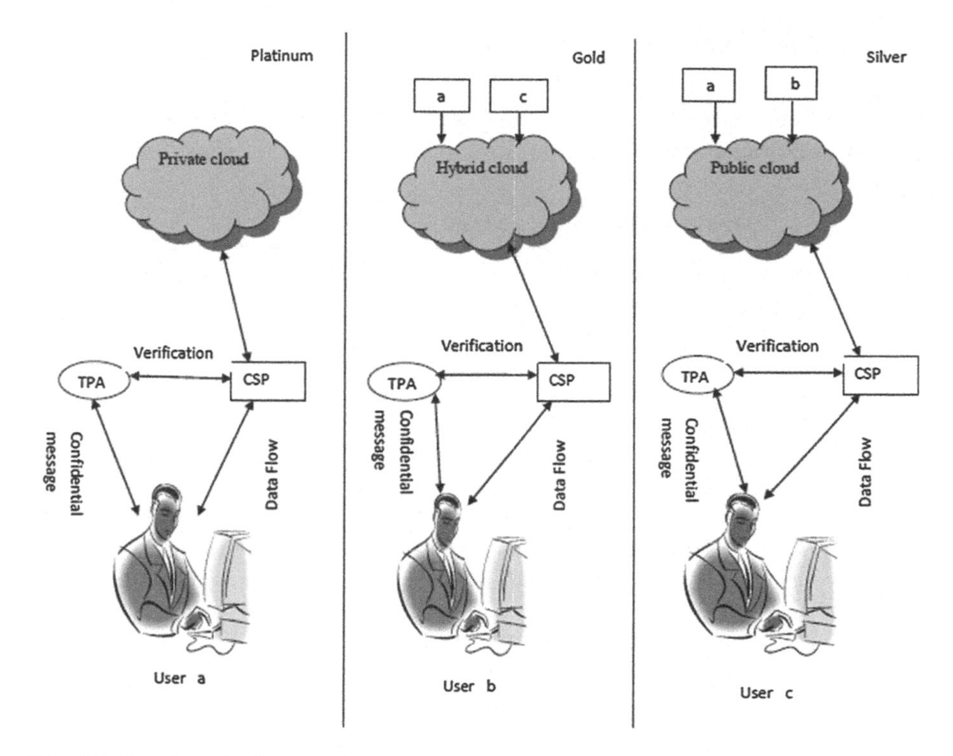

Fig. 6.9 Data integrity framework

It can clearly be seen that depending on type of data, the user can move from one service provider to another, e.g. the user 'a' can put his data over hybrid or public cloud depending on the security needed for the data stored. The same thing can be applied by other two users. Depending on the type of data to be stored on various clouds, there are three main platforms in the model namely:

- *Platinum*—To store sensitive data like data related to transactions of ATMs, bank account information along with high level of security on the data. The data will be stored on a private cloud.
- *Gold*—To store data related to simple login on any page like Facebook, ebooking and email login is stored. The level of security needed is not that high. Security only on password is required.
- *Silver*—To store data related to only simple browsing of sites, uploading of images, downloading of files like downloading of music files or images are stored. The level of security needed is the least.

Based on these levels, the user will decide the platform to store the data.

6.3.1 Data Integrity Algorithms

In this section, we have discussed various algorithms which are used to implement the proposed framework.

Algorithm 6.1: SaaS cloud integrity

```
/* Variables used for:
        User => u,
        Platinum => p,
        Gold => g
        Silver => s
*/
Begin
        If u chooses p
          Then call module m1;
        If u chooses g
          Then call module m2;
        If u chooses s
          Then call module m3;
End

Module m1:
  Begin
     User's data is encrypted using RSA and sent to CSP;
     Data is verified by CSP;
     If data is valid
          success message;
      Go To Module T
  End

Module m2:
  Begin
     Data stored is encrypted using Bcrypt algorithm;
     Data is verified by CSP;
     If data is valid
          success message;
     Go to module T
End

Module m3:
   Begin
     Data is encrypted using AES algorithm;
     Verification of data is done by CSP;
     If data is valid
       success message;
     Go To Module T
    End

Module T:
  Begin
     Check the data stored.
     If  user's data == cloud data then
      data valid;
     Else
       corrupted data;
End
```

In module m1, RSA algorithm [44] is used to provide integrity of data because for storing sensitive information on cloud, hashing algorithms are used. RSA is based on the difficulty of factoring large numbers. There are various advantages of RSA due to which it is preferred over DSA.

- DSA can only be used for authentication while RSA can be used for both authentication and to encrypt a message.
- A bad random number generator will leak DSA key bits.
- Faster at encrypting than DSA.

In module m2, Bcrypt algorithm is used for hashing the passwords over algorithms like MD-5, SHA-1, SHA-2 and SHA-3. A password hashing algorithm should preferably be slow in order to prevent brute force attacks; it should have features which actually decrease the feasibility of a distributed brute force attack on the hashes. Bcrypt algorithm is derived from the Blowfish block cipher which uses lookup tables that are initiated in memory to generate the hash.

In module m3, AES algorithm is used to provide security on the stored data. AES is asymmetric encryption algorithm which is used to encrypt the message. Here, sender uses the public key of receiver and receiver uses its private key to decrypt the message. The following features of AES over DES describe its usability for the framework presented in Fig. 6.10.

- AES is more secure in comparison to DES algorithm.
- AES data encryption is mathematically more efficient and elegant cryptographic algorithm. Key length option is the main strength of the algorithm. Time required to crack an encryption algorithm is directly related to the length of the key used to secure the communication. AES gives an option to choose a 128-bit, 192-bit or 256-bit key, making it exponentially stronger as compared to the 56-bit key of DES.
- Block size of DES is small compared to AES algorithm.
- A balanced Feistel structure is used by DES while substitution-permutation is used by AES.

6.3.2 Performance Analysis of Security Techniques

In order to provide assurance of characteristics in cloud systems like reliability, security, fault tolerance, sustainability and scalability, computational services timely, repeatable and controllable methodologies are required for evaluation of new cloud applications and policies before actual development of cloud products. In [45], Thakur et al. have presented the simulations of above mentioned algorithm using CloudSim. Algorithmic analysis and obtained results using RSA, AES and Bcrypt for the presented framework in Fig. 6.9.

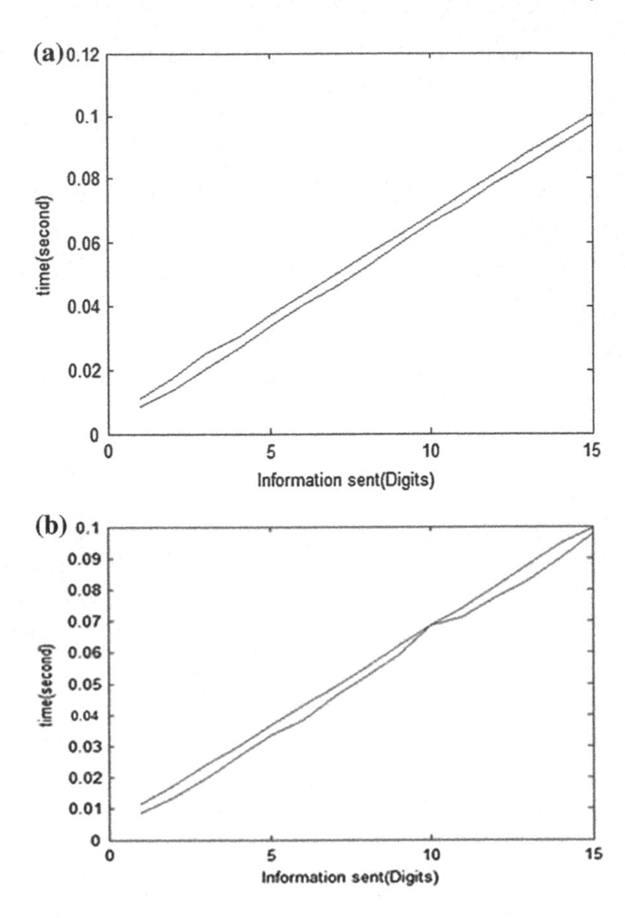

Fig. 6.10 Message length versus time. **a** $p = 3$ and $q = 7$, **b** $p = 23$ and $q = 17$

- *RSA Algorithm Analysis*

RSA algorithm is tested for integer numbers ranging from a single digit message length to 16-digit message length. The execution time t is in seconds. The execution time depends on the values of p and q which are prime numbers. Different values of p and q have been considered and depending on these values, graph between message length and time is plotted (Fig. 6.10a, b).

- *AES Algorithm Analysis*

AES is used here to provide data integrity while simple browsing of webpages over Internet. Plain text is encrypted to hexadecimal format. The change in graph depends on the value of plain text. The time taken increases if there is a use of combination of text and digits. A graph between information sent and time is plotted (Fig. 6.11).

Now by using the discussed framework and the type of information that has to be stored, multilevel security can be provided on different files. Tables 6.5, 6.6 and 6.7 summarize the execution time of four different files using RSA algorithm, Bcrypt algorithm, AES algorithm, respectively.

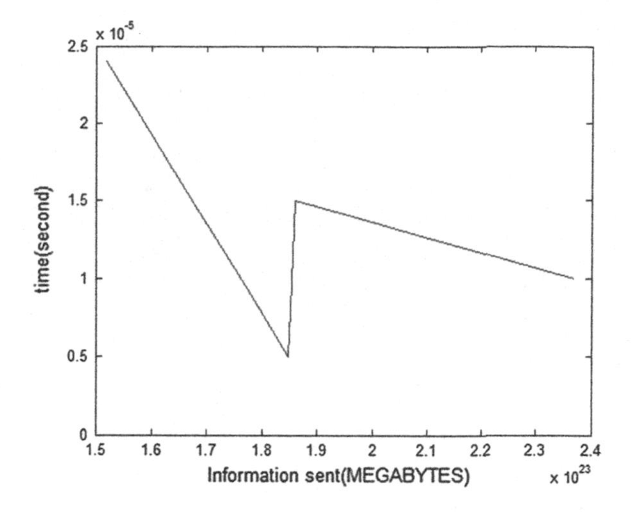

Fig. 6.11 AES algorithm

Table 6.5 File execution time using RSA algorithm

File size (MB)	RSA encryption (ms)	RSA decryption (ms)
1	66,787	720
2.9	2206	385
3.98	17,325	1081
4.65	40,441	1064

Table 6.6 File execution time using Bcrypt algorithm

File size (MB)	BCRYPT encryption (ms)	BCRYPT decryption (ms)
1	1306	447
2.9	1218	454
3.98	1225	510
4.65	1225	445

Table 6.7 File execution time using AES algorithm

File size (MB)	AES encryption (ms)	AES decryption (ms)
1	2129	234,229
2.9	3132	21,663
3.98	3276	71,129
4.65	2613	152,405

Table 6.8 Performance analysis of previous scheme versus proposed framework when number of requests is 8

No. of requests	Previous scheme [21] (time in ms)	Proposed scheme (time in ms)
1	3074	3074
2	3074	2917
3	3075	796
4	3009	1108
5	2995	1108
6	2905	796
7	3079	921
8	3420	827

Table 6.9 Performance analysis of previous scheme versus proposed scheme when number of requests increased to 16

No. of requests	Previous scheme [21] (time in ms)	Proposed scheme (time in ms)
1	3248	3006
2	3291	3110
3	3106	900
4	3341	1249
5	3570	1289
6	3320	1160
7	3334	1276
8	3456	900
9	3418	1000
10	3534	958
11	4018	2987
12	3987	3540
13	3765	3491
14	4211	4010
15	4300	2667
16	4696	3800

To analyse the performance of discussed algorithm, a comparative analysis is also done with previous algorithms using the presented framework. In [21], all files were provided same level of security irrespective of the type of data which is in contrast to the current framework that provides different levels of security on files with varying sizes. Tables 6.8, 6.9 and 6.10 present the comparison between the encryption time of previous scheme and the proposed scheme when the number of requests keeps on increasing. Figure 6.12a–c, represent this process.

Table 6.10 Performance analysis of previous scheme versus proposed scheme when number of requests increased to 32

No. of requests	Previous scheme [21] (time in ms)	Proposed scheme (time in ms)
1	3896	3694
2	3424	3341
3	3308	3208
4	3410	3400
5	3540	3459
6	3691	3576
7	3330	3200
8	3120	3110
9	3600	3567
10	3498	1000
11	3900	3741
12	4009	4003
13	3774	3669
14	4219	4198
15	4166	4047
16	4333	4216
17	3681	3538
18	3908	3497
19	3724	3623
20	3805	3551
21	3500	3348
22	4460	4234
23	4000	3811
24	4790	4626
25	4546	4377
26	3980	3694
27	2196	987
28	2897	1248
29	3071	2381
30	4298	3106
31	4571	3963
32	5168	4685

Tables 6.8, 6.9 and 6.10 show the average time taken for all three cases for proposed framework and performance is evaluated as shown in Table 6.11.

The average time taken by the schemes in both the cases is shown in Fig. 6.13. It can be predicted easily that the time taken by proposed scheme is less than the time taken by previous scheme.

In case, if the file is modified then the data in file will get encrypted using RSA algorithm but it will not be decrypted as shown in Table 6.12.

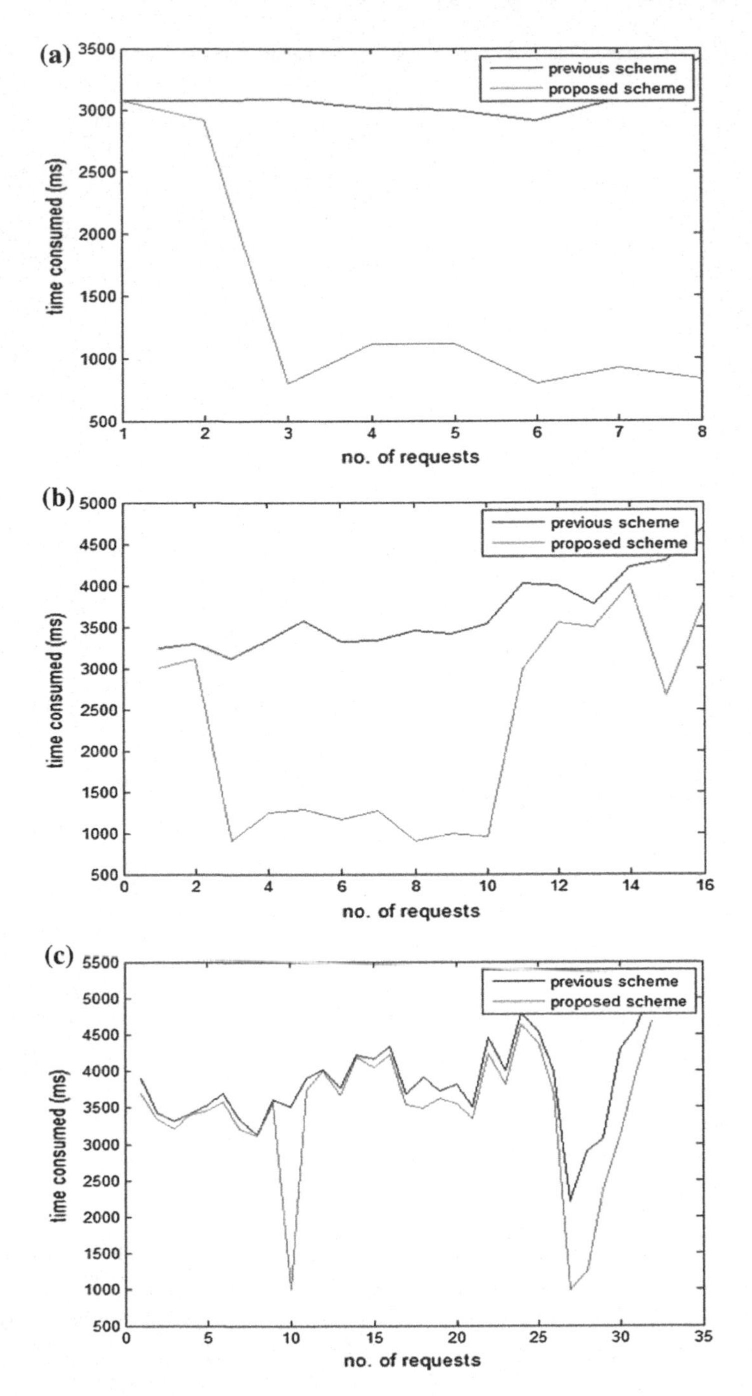

Fig. 6.12 Encryption time for previous scheme and proposed scheme: **a** number of requests = 8, **b** number of requests = 16, **c** number of requests = 32

Table 6.11 Average time taken by previous schemes and proposed scheme when requests vary from 8 to 32

No. of requests	Previous scheme [21] Avg. time (ms)	Proposed scheme Avg. time (ms)
8	3105.91	1086.3
16	3870.19	2519.88
32	3899.72	3149.19

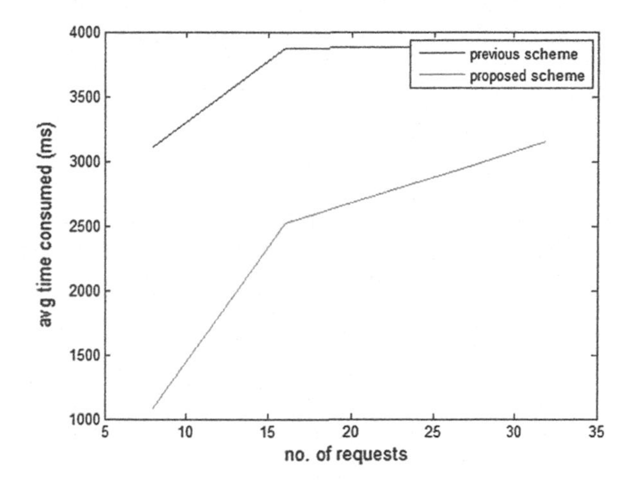

Fig. 6.13 Average time taken by both the schemes

Table 6.12 Encryption and decryption time when file is modified

File size (MB)	Algorithm	Encryption (ms)	Decryption (ms)
1	AES	1037	3184
2.90	Bcrypt	698	1786
3.98	RSA	1123	–
4.65	Bcrypt	695	1781

6.4 Summary

Cloud computing is becoming one of the crucial technologies and considered as a boon for the IT sector. It follows multi-tenancy where several users can share the resources at a given time over the Internet. Along with the benefits of cloud computing there are some drawbacks, in which security is major concern which causes hinderance in full acceptance of this technology. In this chapter, we have summarized various security issues and security threats related to each level of cloud. We have proposed a framework to maintain data integrity in SaaS. This framework also implements the various security algorithms at three different levels. The performance analysis of this framework represents that the proposed framework is easy to implement and can effectively maintain the information security in clouds.

References

1. Shaikh FB, Haider S (2011) Security threats in cloud computing. In: Proceeding of international conference for internet technology and secured transactions (ICITST). IEEE, pp 214–219
2. Alassafi MO, Hussain RK, Ghashgari G, Walters RJ, Wills GB (2017) Security in organisations: governance, risks and vulnerabilities in moving to the cloud. Enterp Secur 241–258
3. Tianfield H (2012) Security issues in cloud computing. In: Proceedings of IEEE international conference on systems, man, and cybernetics (SMC'12), Seoul, Korea, pp 1082–1089
4. Gugnani G, Ghrera SP, Gupta PK, Malekian R, and Maharaj BTJ (2016) Implementing DNA encryption technique in web services to embed confidentiality in cloud. In: Proceedings of the 2nd international conference on computer and communication technologies. Springer, pp 407–415
5. Magoulès F, Pan J, Teng F (2012) Cloud computing data-intensive computing and scheduling. CRC Press, Boca Raton, p 231
6. Naruchitparames J, Güneş MH (2011) Enhancing data privacy and integrity in the cloud. In: Proceedings of 2011 international conference on high performance computing and simulation (HPCS), pp 427–434
7. Aluvalu R, Muddana L (2015) A survey on access control models in cloud computing. In: Proceedings of 49th annual convention of the Computer Society of India (CSI) Emerging ICT for bridging the future, 1, pp 653–664
8. Rashdi A et al (2013) Cloud security standards. Oman National CERT Information Technology Authority, pp 1–28
9. Sun D, Chang G, Sun L, Wang X (2011) Surveying and analyzing security, privacy and trust issues in cloud computing environments. Proced Eng 15:2852–2856
10. Xiao Z, Xiao Y (2013) Security and privacy in cloud computing. IEEE Commun Surv Tutor 15(2):843–859
11. Thakur AS, Gupta PK (2014) Framework to improve data integrity in multi cloud environment. Int J Comput Appl 87(10):28–32
12. Zissis D, Lekkas D (2012) Addressing cloud computing security issues. Future Gener Comput Syst 28(3):583–592
13. Guilloteau S, Venkatesen M (2012) Privacy in cloud computing. ITU-T technology watch report, pp 1–26
14. Chen D, Zhao H (2012) Data security and privacy protection issues in cloud computing. In: Proceedings of IEEE-conference on computer science and electronic engineering, pp 647–651
15. Sun Y, Zhang J, Xiong Y, Zhu G (2014) Data security and privacy in cloud computing. Int J Distrib Sens Netw 1–9
16. Pearson S (2013) Privacy, security and trust in cloud computing. Privacy and security for cloud computing, pp 3–42
17. Shin D, Ahn GJ (2005) Role-based privilege and trust management. Comput Syst Sci Eng J 20(6):401–410
18. Rana P, Gupta P, Siddavatam R (2014) Combined and improved framework of infrastructure as a service and platform as a service in cloud computing. In: Babu B et al (eds) Proceedings of the 2 nd international conference on soft computing for problem solving (SocProS 2012), December 28–30, 2012. Adv Intell Syst Comput 236:831–839
19. Eswaran S, Abburu S (2012) Identifying data integrity in the cloud storage. IJCSI Int J Comput Sci Issue 9(2):403–408
20. Joshi BK, Shrivastava MK, Joshi B (2016) Security threats and their mitigation in infrastructure as a service. Perspect Sci 8:462–464
21. Mohta A, Sahu RK, Awasthi LK (2012) Robust data security for cloud while using third party auditor. Int J Adv Res in Comput Sci Software Eng 2(2):1–5

22. Subashini S, Kavitha V (2011) A survey on security issues in service delivery models of cloud computing. J Netw Comput Appl 34(1):1–11
23. da Silva CMR, da Silva JLC, Rodrigues RB, Nascimento LMD, Garcia VC (2013) Systematic mapping study on security threats in cloud computing. Int J Comput Sci Inf Secur 11(3):1–10
24. Islam T, Manivannan D, Zeadally S (2016) A classification and characterization of security threats in cloud computing. Int J Next Gener Comput 7(1):1–20
25. Vaquero LM, Rodero-Merino L, Morán D (2011) Locking the sky: a survey on IaaS cloud security. Computing 91(1):93–118
26. Hashizume K, Rosado DG, Fernández-Medina E, Fernandez EB (2013) An analysis of security issues for cloud computing. J Internet Serv Appl 4(1):1–13
27. Kazim M, Zhu SY (2015) A survey on top security threats in cloud computing. IJACSA 6 (3):109–113
28. TTW Group et al (2013) The notorious nine: cloud computing top threats in 2013. Cloud Security Alliance
29. Chou TS (2013) Security threats on cloud computing vulnerabilities. IJCSIT 5(3):79–88
30. Dillon T, Wu C, Chang E (2010) Cloud computing: issues and challenges. In: Proceedings of 24th IEEE international conference on advanced information networking and applications, pp 27–33
31. Yan L, Rong C, Zhao G (2009) Strengthen cloud computing security with federal identity management using hierarchical identity-based cryptography. In: Proceedings of international conference on cloud computing, IEEE, pp 167–177
32. McIntosh M, Austel P (2005) XML signature element wrapping attacks and countermeasures. In: Proceedings of workshop on secure web services.ACM, New York, USA, pp 20–27
33. Constantin L (2011) Researchers demo cloud security issue with AMAZON AWS attack. IDG News Service, Available via PCWorld. https://www.pcworld.idg.com.au/article/405419/researchers_demo_cloud_security_issue_amazon_aws_attack/. Accessed 15 Feb 2017
34. Mushtaq MO, Shahzad F, Tariq MO, Riaz M, Majeed B (2017) An efficient framework for information security in cloud computing using auditing algorithm shell (AAS). IJCSIS 14 (11):317–331
35. Kritikos K, Massonet P (2016) An integrated meta-model for cloud application security modeling. Proced Comput Sci 97:84–93
36. Nafi KW, Kar TS, Hoque SA, Hashem MMA (2012) A newer user authentication, file encryption and distributed server based cloud computing security architecture. Int J Adv Comput Sci Appl 3(10):181–186
37. Metri P, Sarote G (2011) Privacy issues and challenges in cloud computing. Int J Adv Eng Sci Technol 1(5):1–6
38. Mathew A (2012) Security and privacy issues of cloud computing; solutions and secure framework. Int J Multidiscip Res 2(4):182–193
39. Gajanayake R, Iannella R, Sahama T (2011) Sharing with care an information accountability perspective. Internet Comput 15:31–38
40. Chidambaram N, Raj P, Thenmozhi K, Amirtharajan R (2016) Enhancing the security of customer data in cloud environments using a novel digital fingerprinting technique. Int J Dig Multimed Broadcast 1–6
41. Hamlen K, Kantarcioglu M, Khan L, Thuraisingham B (2010) Security issues for cloud computing. Int J Inf Secur Priv 4(2):39–51
42. Juels A, Kaliski Jr BS (2007) PORs: proofs of retrievability for large files. In: Proceedings of 14th ACM conference on computer and communications security. ACM, pp 584–597
43. Chalse R, Selokar A, Katara A (2013) A new technique of data integrity for analysis of the cloud computing security. In: Proceedings of 5th IEEE international conference on computational intelligence and communication networks, pp 469–473
44. Singh S, Maakar SK, Kumar DS (2013) A performance analysis of DES and RSA cryptography. IJETTCS 2(3):418–423
45. Thakur AS, Gupta PK, Gupta P (2014) Handling data integrity issue in SaaS cloud. In: Proceedings of FICTA, Springer. pp 127–134

Chapter 7
Applications of Predictive Computing

7.1 Introduction

Predictive computing is used in a wide spectrum of real-world problems ranging from business, government, economics and also science. Some major fields reaping the advantages of predictive computing are financial services, insurance, telecommunications, retail, healthcare and government [1]. Predictive computing is now adopted by almost all major work sectors as its advantages are evident in every field. Predictive computing helps in knowing the customers and acts on the insights provided by the predictive analytics on future customer behaviour. This can help to identify the best actions to take for every customer or transaction [2] and help in guiding other strategic actions to be taken for the profit of business, such as collaborating with agencies that maximize the profit in the approved budget, or detecting frauds or abuse in insurance or healthcare claims. It can help in answering complex questions such as live transactions, with empirical precision at incredible speeds [2]. Decisions that earlier took hours or days, can now be taken in milliseconds. Insights given by analytics can be helpful as they can reduce the business losses by accurately measuring the risks and frauds. Predictive computing can detect the slightest abnormality in a pattern of usual business routine transaction or data, and hence can help reduce business losses. Predictive computing adds consistency and stability to business decisions and in turn improves customer service as it relies on mathematical models and techniques. Also, the decisions provided by predictive analytics are consistent and unbiased as compared to human experts. Predictive computing is a requirement of today's world as everyone needs consistent, faster, smarter decisions to meet the agile business standards, where market conditions change quickly and frequently. Also, it is required to improve the customer service and grow the profit of the business. Predictive computing improves every aspect of decision-making process, which includes: precision, consistency, agility, speed and cost [2]. There are uncountable applications of predictive computing in business intelligence, and this is because of the fact that predictive

P.K. Gupta et al., *Predictive Computing and Information Security*,
DOI 10.1007/978-981-10-5107-4_7

computing has reached the roots of various business sectors. Predictive computing allows us to anticipate the future and make an optimal decision by extracting information from the datasets, which in turn helps to discover complex relationships, recognize unknown patterns, forecasting actual trends, etc. [3].

7.2 Applications Based Features of Predictive Computing

Considering the various horizons for predictive computing as presented in Table 1.2 of Chap. 1, we have selected three application-based scenarios, i.e., smart mobility, e-Health and e-Logistics to be presented in this chapter. Each scenario consists of various applications, where predictive computing can be successfully implemented. Following sections discuss about these applications.

7.2.1 *Smart Mobility*

Urbanization of the cities of developing countries has led to a steady rise in the number of vehicle registrations, traffic congestion problems, heavy fuel demands and fuel consumption, and financial and economic challenges. In the past several years, smart mobility (e-mobility) has been a subject of intensive research and discussion revolving around ecological and economic arguments around the world [4]. Among the varied discussions, countries like Singapore, China, Japan, India and South Korea, have presented interesting approaches while promoting and implementing smart mobility. Key factors driving smart mobility among others are smart vehicle navigation, finding optimized shortest paths, smart technological capabilities of the vehicle to reduce fuel consumption. Smart mobility can be sought out as a solution to congestion, increasing air pollution, noise pollution in the cities of developing countries. Smart mobility can be considered for addressing these problems as well. In [5], Pattanaik et al. have proposed a *smart congestion avoidance technique* by estimating the scope of real-time traffic congestion on urban road networks which also *predicts an alternate shortest route to the destination* as shown in Algorithm 7.1. The proposed system uses *k-Means clustering algorithm* to estimate the magnitude of congestion on different roads, later it employs *minimum spanning tree algorithm; Dijkstra's algorithm* to predict the shortest route. Once the system receives the user's destination coordinates, it predicts the shortest route from the user's current location. This process is reiterated until the user reaches the desired destination. The proposed methodology can predict which road segments are congested or clear through the real-time GPS data and inform the user about real-time traffic conditions and adjusts the route so as to avoid congestions and reduce travelling time. The congestion avoidance algorithm is given below.

Algorithm 7.1: Congestion Avoidance Algorithm [5]

1. Fetch Driver's Current Location (Source)
2. Get Destination Coordinates
3. Retrieve Road Map of Area
4. **while** *Source* $\neq$ *Destination* **do**
5. Retrieve Real-Time Traffic Data from App
6. Plot Traffic Data onto 2D Problem Space
7. Apply *k-Means Clustering* on Traffic Data
8. Assign Weights to Traffic Clusters
9. Combine Traffic Cluster Data with Road Map
10. Convert Weighted Road Map into Neighborhood Matrix
11. Apply *Dijkstra's Algorithm*
12. Display Shortest Path
13. Fetch Driver's Current Location (Source) Again
14. **end while**

In another work, Milojevic and Rakocevic [6] have presented that vehicular traffic congestion is becoming a major economic and social problem which requires the government's utmost attention. It leads to significant financial and safety challenges, and also a major contributor to the increasing pollution in the cities. They have proposed a new vehicle-to-vehicle (V2V) congestion detection algorithm based on the IEEE 802.11p standard. This algorithm allows vehicles to be self-aware about road conditions and finds congestion detection based on the monitoring of speed and cooperation with the surrounding vehicles. Proposed algorithm comprises of five steps as shown in Algorithm 7.2, (1) *Speed Monitoring*, (2) *Congestion Detection*, (3) *Localization*, (4) *Aggregation* and (5) *Broadcasting*. They have also presented a performance evaluation using large-scale simulation in Veins framework based on the OMNet++ network simulator and SUMO vehicular mobility simulator. Results show that precise congestion detection and qualification can be achieved using a significantly decreased number of exchanged packets.

Algorithm 7.2: Congestion Control Algorithm [6]

1) Speed Monitoring
IF $V_c \neq V_t$ **GOTO** 2;
2) Congestion Detection
IF $V_c < V_t$ **THEN**
(Start Timer τ_c, WHEN $\tau_c = \eta.10s \Rightarrow C_p = \eta$);
ELSE (Start Timer τ_c, WHEN $\tau_c = 10s \Rightarrow C_p = 0$);
3) Localization
FIND A_{id} of the current location, **GOTO** 4;
4) Aggregation
GET $C_d(A_{id})$;
IF ($C_p \neq 0$) **THEN**
IF $C_p(A_{id}) > C_d(A_{id})$ **THEN** 5, $C_d(A_{id}) = C_p(A_{id})$;
ELSE skip 5;
ELSE IF $C_p(A_{id}) \neq C_d(A_{id})$ **THEN** 5, $C_d(A_{id}) = C_p(A_{id})$ **THEN** 5;
ELSE skip 5;
5) Broadcasting
Broadcast the (C_p, A_{id});

In [7], Abhishek et al. have presented a study to analyse and resolve the congestion of the complex traffic conditions in the cities. Proposed algorithm tries to control and optimise the duration of time for which the traffic light signal is green, and the number of vehicles passing through the junction during that time period. Wireless sensor network has been used to make the traffic signals adaptive to the dynamic traffic flow, so that the number of vehicles passing through the signal is maximized. Following parameters have been considered while the development of proposed algorithm: (1) *Waiting Time*, (2) *Clearance Time*, (3) *Rate of Arrival*, (4) *Proportionality Constant*, (5) *Clear Route* and (6) *Multiplication Factor*. In [8], a new smart traffic control design is presented which resolves traffic issues and utilizes available road infrastructure. Authors have also considered reducing the waiting time, fuel consumption, traffic congestion and levels of traffic obstruction. Intelligent Traffic Control System (ITSC) is based on a principle stated as 'a car can only move ahead if there is space for it to move ahead' and 'the signal remains green until the present cars have passed'. Here, sensors are placed at every entry and exit of a junction, and are responsible for monitoring the number of cars present at the junction to make traffic management smooth and efficient. Ye et al. [9] and Malekian et al. [10] have proposed driving route prediction methods based on the *Hidden Markov Model* (*HMM*). This method can accurately predict a vehicle's entire route as early in a trip's lifetime as possible without inputting origins and destinations beforehand. First, the driving route recommendation system architecture is proposed which highlights a method for route prediction based on the

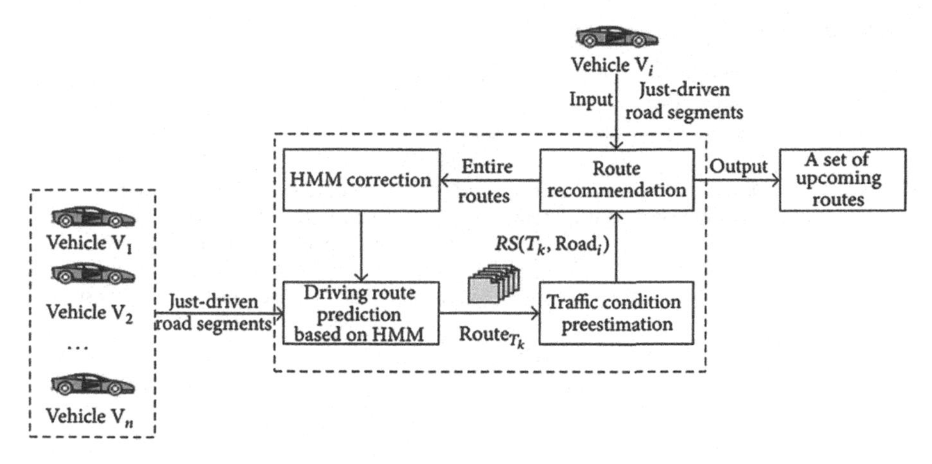

Fig. 7.1 Architecture of route recommendation system [9]

knowledge of HMM. The method can predict the congested road segments as well as smooth road segments through route prediction. The system also updates traffic information in real time and informs the driver to adjust the driving route as the trip progresses. Figure 7.1 shows the architecture of the proposed driving route recommendation system.

This architecture consists of four phases; (1) *Driving Route Predictions Based on HMM*, (2) *Traffic Congestion Pre-estimation*, (3) *Vehicle Route Recommendation* and (4) *HMM Correction* for route prediction and recommendation. Table 7.1 summarizes various applications designed and used for smart mobility of vehicles.

7.2.2 e-Health

Healthcare is another major sector that has witnessed the wide use of wireless sensor networks, IoT, and Cloud computing to perform various types of predictions related to patient's health, monitoring of blood pressure, heart beats, breast cancer, lung disease, etc. Diagnosing from early symptoms or patterns, predictive computing can be used at each level of e-Health-related applications. Health applications designed and used previously, have been shifted to e-Health [30]. Earlier, medical applications were based on the analog telephony that enables the individuals to call the healthcare professional, hospitals to take appointments, and to transmit electrocardiograms over telephone lines [30]. Tele-Health involves health services delivered from a distance and is an important constituent of e-Health [31], but these analog techniques could not be expanded due to the bandwidth limitations, low rate of data transfer over copper wires, the existence of inference and noise. Another constituent of e-Health is 'm-health' which can be defined as a medical and public health practice supported by mobile devices, such as mobile phones, patients monitoring devices and other wireless devices [32].

Table 7.1 Applications for smart mobility

S. no.	Application	Purpose
1	Smart real-time traffic congestion estimation [5]	An intelligent system that predicts congested road segments through real-time GPS data, and suggests shortest alternate route
2	Distributed vehicular traffic congestion detection algorithm [6]	A system that uses a new vehicle-to-vehicle (V2V) congestion detection algorithm which allows the vehicle to be smart. It allows vehicles to be self-aware of the traffic on the road based on the speed analysis and assistance with neighbouring vehicles
3	City traffic congestion control [7]	A wireless sensor network used to analyze and resolve the congestion of the complex traffic conditions in the cities
4	Smart navigation for visually impaired [11–14]	An intelligent, portable, and user-friendly system that enables the visually impaired people to navigate in an environment without any human intervention. It provides assistance for both indoor and outdoor navigation by voice input and output
5	Modelling urban traffic dynamics [15]	A smart system that proposes a way to capture the complexity based on generalization power of Markov chains in coexistence with continuous urban data streams
6	Remote monitoring of vehicle diagnostics [16]	A distributed system that is used for remote monitoring of vehicles and their geographical position using a smart box with GPS and GPRS
7	Intelligent traffic control system [8, 17]	A smart traffic control design that resolves traffic issues and utilizes 100% use of the road infrastructure
8	Route predictions based on hidden Markov model [9]	A system based on the HMM that can accurately predict a vehicle's entire route at the earliest in a trip's lifetime
9	Vehicle tracking system using IoT [18–20]	An intelligent system that makes use of built-in internet enabled GPS sensor to render a real-time and reliable vehicle tracking services to public
10	Intelligent ambulance using IoT [21]	An intelligent system that proposes to make way for ambulances in heavy traffic congestion traffic situations by manipulating the timer of the signal light board
11	Smart parking system using IoT [22–25]	A smart parking system that enables the user to identify the closest parking area and provides the availability of vacant parking slot in the current or neighbouring area
12	Improvement of traffic monitoring system using IoT [26]	An adaptive traffic congestion control system that provides time slot to each route based on the traffic density by fetching the location and

(continued)

Table 7.1 (continued)

S. no.	Application	Purpose
		speed of the vehicle using GPS, which in turns enables us to calculate the accurate time when the vehicle reaches the next intersection
13	IoT-RFID testbed for supporting traffic light control [27]	An IoT-based application for traffic management system that enables the traffic police officers in support decision making
14	Framework for agent-based traffic light control [28]	A smart agent-based framework that proposes a balanced, coordinated and optimal method for traffic light control
15	Real time traffic congestion detection system [29]	An intelligent system that detects real-time congestion without any human intervention or human supervision without any prior knowledge about the condition

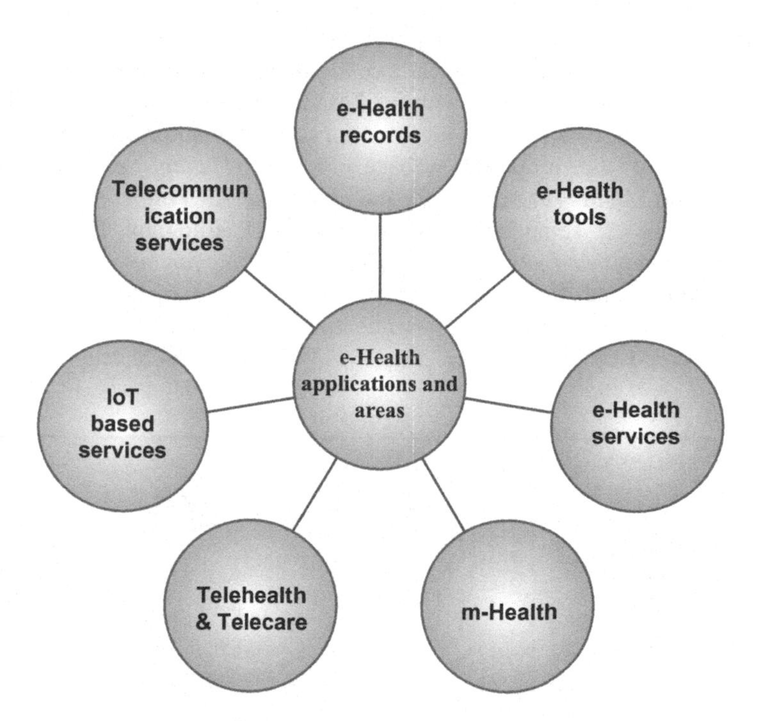

Fig. 7.2 e-Health applications and its target areas

e-Health has become part of every citizen's everyday lives and impacts them in one way or other, and this has led to the development of various e-Health related applications. e-Health applications use the information and communication technologies (ICT) for handling various health-related services. It deals with the broad spectrum of e-Health policies, legal and ethical frameworks, adequate funding and training [33]. The target areas of e-Health are depicted in Fig. 7.2. It has seen a tremendous growth over the past 30 years, enabling the exchange of healthcare and administrative data and transfer of medical images and laboratory results [30].

In [34], Gupta et al. have discussed an IoT-based cloud-centric healthcare architecture predictive analysis of physical activities of the users in sustainable health centers. The prerequisite of this framework is that health centres should be well equipped with sensors for capturing the patient's basic health parameters while exercising, such as heart rate, distance, speed, and calories burned daily by a user. These parameter values are stored at the end of the session using either a public or private cloud. Next, the healthcare personnel can access this stored information when required. An alert is sent automatically to the healthcare personnel, if any irregularity is predicted in the user's activity or basic parameters, and an action is initiated by the healthcare personnel. In [35], Baccar and Bouallegue have proposed a novel website architecture for an e-Health program based on a wireless sensor network. Designed website offers an ergonomic and multifunctional platform for an intelligent hospital. Features of this website include management of patient's records, real-time monitoring of patient's condition and geo localization for patients as well as professionals involved with the hospital. The system shown in Fig. 7.3, uses remote sensing of biometrics signals for patient's monitoring. The main three functionalities of the proposed website are; (1) *Manage patients records*: Add/Delete and Modify diagnostics for the health file, (2) *Follow the vital signals progress of patients:* Temperature, Blood Pressure, Cardiograph Pulses, etc. and (3) *Localize patients and professionals; mapping service for out-patients.*

In [36], Ahmed and Abdullah have presented an e-Health model from ubiquitous perspective. This model provides data acquisition, archiving, and presentation in the cloud. The proposed model makes use of cloud service architecture (CSA) for

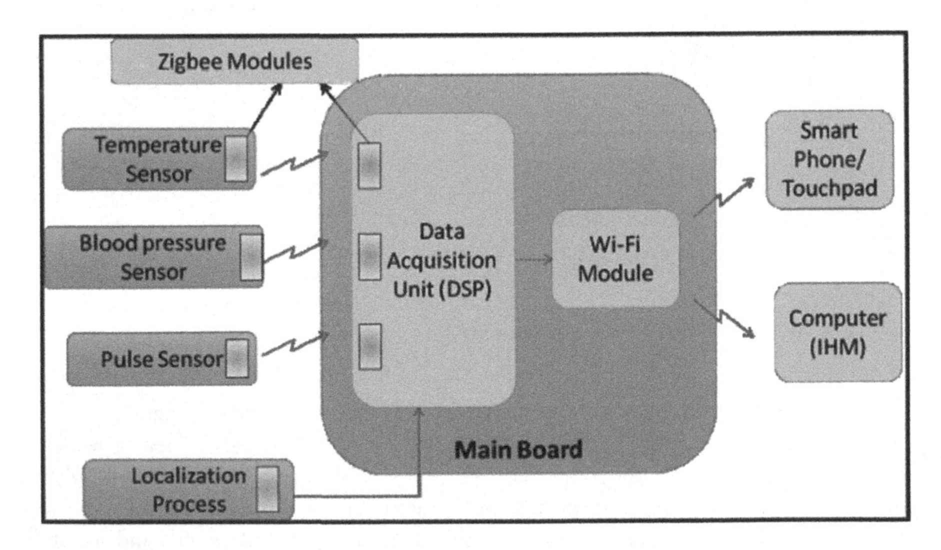

Fig. 7.3 Main board for e-Health platform [35]

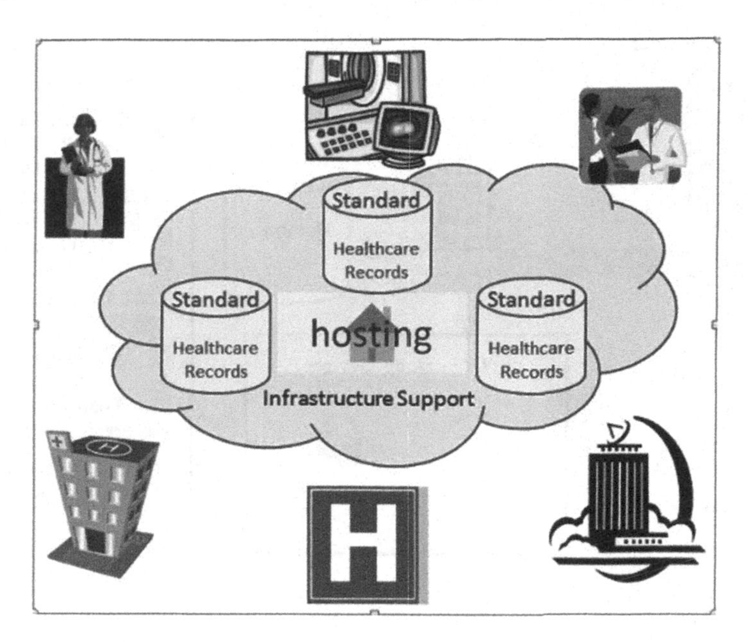

Fig. 7.4 e-Healthcare cloud [37]

processing of medical information of a patient. In another work [37], Liu and Park have focused on the e-Healthcare application cloud-enabling characteristics. The authors of this research work found close proximity of the proposed e-healthcare architecture and the cloud environment. The e-healthcare cloud is shown in Fig. 7.4.

Authors have also discussed the challenges in the adaptation of a pure cloud solution for smart e-healthcare. In [38], Aruna et al. have designed a patient health monitoring system (PHMS) which includes three phases; (1) *collection phase*, (2) *transmission phase* and (3) *utilization phase*. A Body Area Network (BAN) is constructed and used to collect the required data from the patient. PHMS notifies the registered patient with the possible precautionary measures to be carried. It suggests the patient with medical care and further steps to be followed in case of critical conditions. A typical architecture of PHMS is shown in Fig. 7.5.

In [39], Piliouras et al. implemented the Electronic Health Records (EHR) technologies. They have listed various challenges that were experienced while integrating EHR technology within the workflow of an already existing healthcare setting. Authors have also listed the various lessons learned from its implementation: (1) *Identify System Champions*, (2) *Give users a lot of Training*, (3) *Perform Root Cause Analysis* and (4) *Quality Management Principles*. Table 7.2 summarizes various applications designed and implemented for e-Health of user or patient.

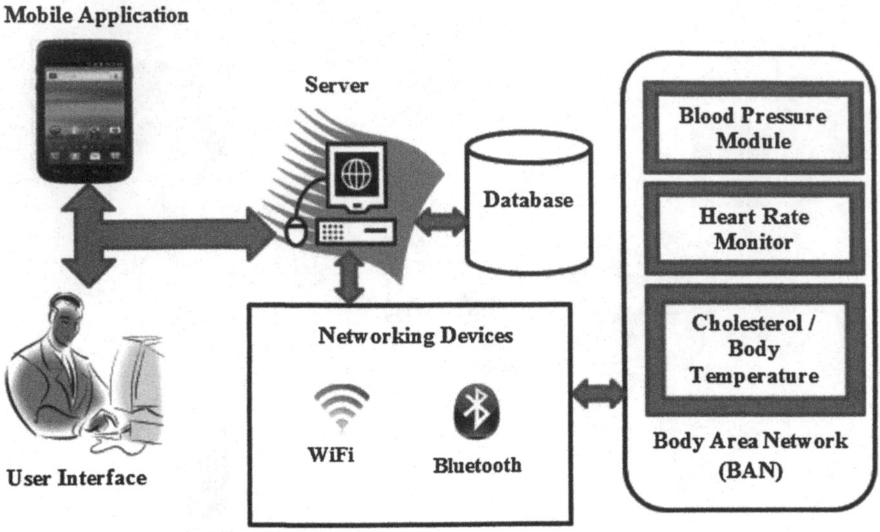

Fig. 7.5 Architecture of PHMS [38]

Table 7.2 Applications for e-Health

S. no.	Application	Purpose
1	IoT-based cloud-centric architecture to perform predictive analysis of user activities in sustainable health centres [34]	A cloud-centric architecture using IoT that uses predictive analysis of the physical activities of the users in maintainable health centres
2	Web-based e-Health platform [35]	A web-based architecture for an e-Health program using a wireless sensor network that offers an ergonomic and multifunctional platform for a smart hospital
3	e-Healthcare management system-based on the cloud-and service-oriented architecture [40]	A smart, and flexible e-Health management system based on cloud computing and SOA and using RIA that overcomes various shortcomings in existing healthcare management systems
4	e-Health monitoring architecture [41]	An intelligent e-Health management system that supports real-time analysis of various parameters of patients using smart devices and wireless sensor networks
5	e-healthcare and data management services in a cloud [36]	An e-Health model that addresses the issues regarding management and security concerns in cloud domain. It includes wireless sensor networks, communication, and storage systems for any hospital using CSA

(continued)

Table 7.2 (continued)

S. no.	Application	Purpose
6	Healthcare cloud computing application solutions [37]	An intelligent cloud-based system that focuses on the cloud-enabling characteristics of e-Healthcare applications
7	Health monitoring and management using internet-of-things (IoT) [42, 43]	An IoT-cloud-based remote patient monitoring system that is used for health monitoring and management
8	Smart real-time healthcare monitoring and tracking System [44]	A smart real-time healthcare monitoring and tracking system that aims to bridge the gap between patients and healthcare professionals
9	IoT-based smart healthcare kit [45]	An intelligent and robust health monitoring system that is capable of monitoring the patient automatically using IoT
10	Remote mobile health monitoring system [46]	A remote mobile health monitoring system that uses mobile and web service capabilities to delivers an end-to-end solution to health-related queries
11	IoT-based remote health monitoring system [47]	An IoT-based health monitoring system that addresses a key challenge to efficiently transmitting healthcare data within the limits of the existing network infrastructure
12	Patient health monitoring system (PHMS) using IoT [38, 48, 49]	An intelligent PHMS that collects patient's sample values and suggests the patient with medical care and further steps to be followed in case of critical conditions
13	Smart cards in healthcare information systems [50–52]	A smart card in healthcare that enables access to a patient's health record stored on network
14	Electronic health record systems [39]	An electronic health record system that uses EHR technologies
15	e-Health tele-media application for patient management [53]	An intelligent e-Health system that provides patients with real-time management of illness, reaction/side effects to prescribed medication, prescription reminders, provides survey data, etc.

7.2.3 e-Logistics

The logistics business has changed dramatically over last few years. Today, the differentiators are more strategic: benchmarking, innovation, network modelling, etc. Employing suitable logistics strategies is a necessity. However, logistic strategy planning is becoming more and more challenging due to dynamically changing scenarios and difficulties in integrating information from different partners. Information from various sources could be combined to generate integrated knowledge that could support the planning process. Integrated knowledge can better describe the potentials of synergy between the available sources of information, and accordingly better exploit logistics strategy planning. In the following proposal, a machine learning-based adaptive framework for logistics planning is proposed. The proposed system will evolve, adapt and improve as its knowledge grows providing a generalized solution to all kinds of logistics activities.

In [54], Khayyat and Awasthi have conducted a study that investigates the problem of collaboration planning in logistics and proposed an agent-based approach for better management of collaborative logistics. Based on the approach, they have designed a support system which utilizes RFID technology for ensuring inventory accuracy. The proposed approach involves three steps: (1) a conceptual agent-based model is designed, (2) the game theory method is utilized to intensively study and analyse suppliers' collaboration and carriers' collaboration that represent major objectives proposed in the preceding model, (3) correctness of the games is verified by formulating them mathematically. Figure 7.6 shows the design of the conceptual multi-agent-based model.

In [55], Wrighton and Reiter have discussed the problem faced in urban cities of Europe, i.e. the transport of goods contributes to the adverse condition of

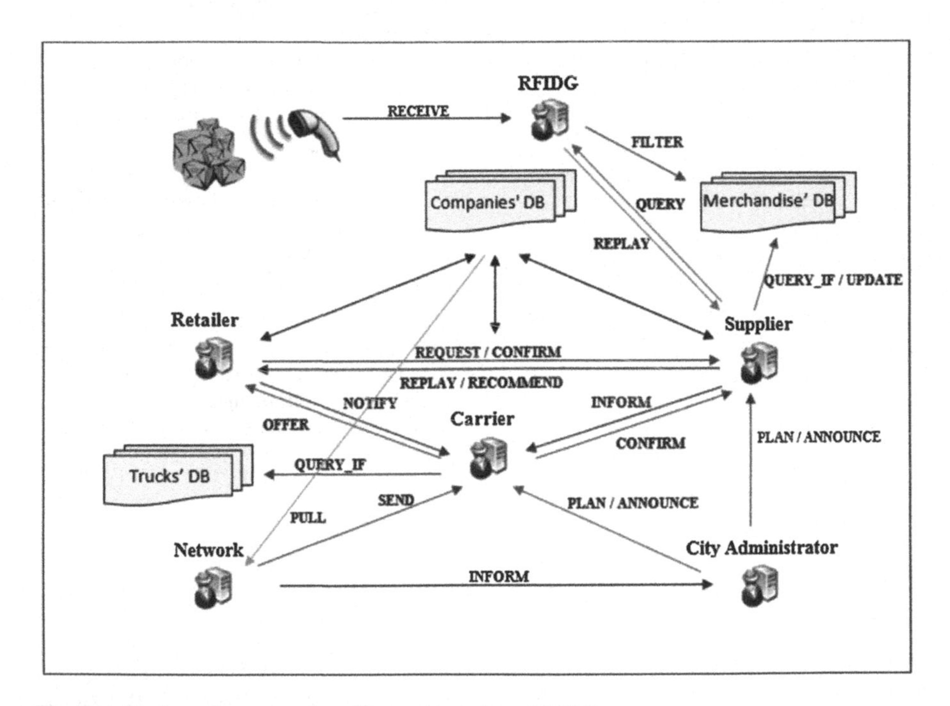

Fig. 7.6 Design of conceptual multi-agent-based model [54]

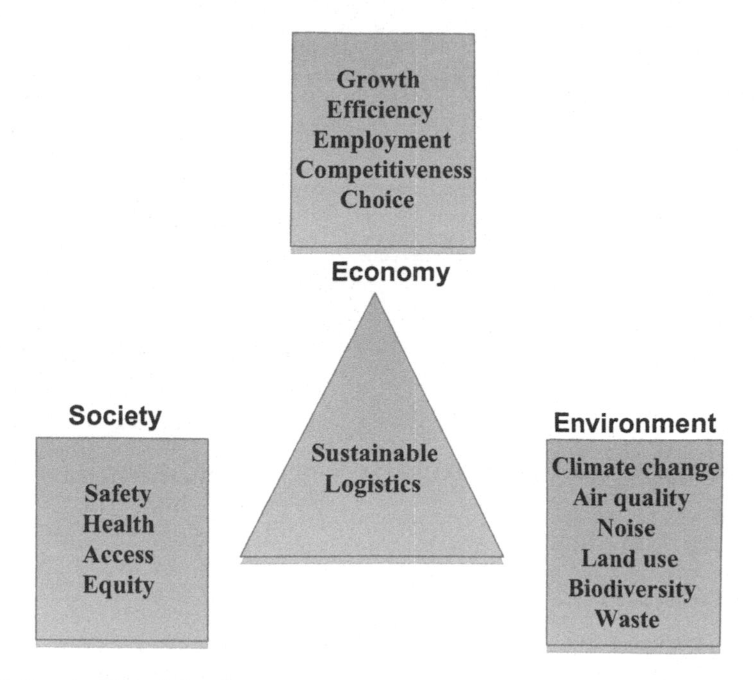

Fig. 7.7 Sustainable logistics dimensions [56]

overcrowding by motorized traffic. City administrators of Europe are aware of the fact that if early measures are not taken to improve this scenario then this will result into a problematic situation. The Cyclelogistics (2011–2014) and Cyclelogistics (2014–2017) projects offer a possible solution to the stated problem. Authors have demonstrated the great potential for the reduction in energy consumption and pollutants caused by urban goods transport by shifting intra-urban final delivery of goods from the car to the bicycle. In [56], the main objective of proposed work is to suggest Smart City logistics on the cloud computing model. Authors have discussed the smart city logistics in terms of sustainable logistics dimensions: Economy, Society and Environment, as shown in Fig. 7.7.

Due to many beneficial characteristics of cloud, the smart logistics has been shifted to cloud computing. Cloud implies a broad range of benefits to the enterprise and other organizations. In [57], the authors have proposed a smart logistics vehicle management system based on Internet of Vehicle (IoV). IoV for smart logistics vehicle management provides various services such as; (1) fleet management, (2) smart driving and (3) transport management. The proposed smart logistics vehicle management consists of following modules based upon the functionality: (1) data collection module, (2) communication module and (3) computational module. In [58], Jianyu and Runtong have represented the model to resolve physical distribution and its effects on E-commerce. Physical distribution is a bottleneck in E-commerce. Authors have constructed the distribution system in

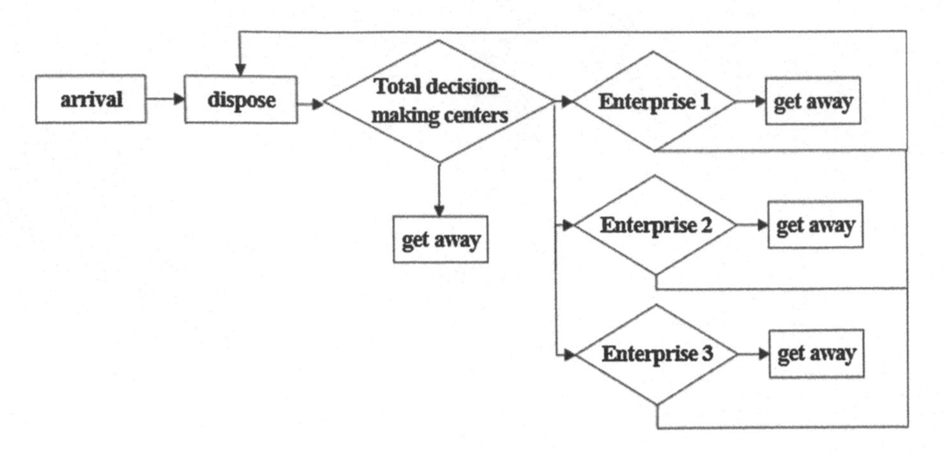

Fig. 7.8 Non-grid simulation logic model diagram about ELS [58]

Table 7.3 Applications for e-Logistics

S. no.	Application	Purpose
1	A comprehensive framework for measuring performance in a third-party logistics provider [59]	A smart framework that provides comprehensive and innovative performance measurement framework for a third-party logistics provider
2	An intelligent multi-agent-based model for collaborative logistics systems [54]	An intelligent multi-agent-based model that has proposed and agent-based approach for better management of collaborative logistics
3	Automobile logistics service supply chain based on reliability [60]	A smart system that deals with building a two-stage automobile LSSC, where the dominant LSI integrates FLSP
4	Cyber-physical logistic systems [61]	An intelligent cyber-physical logistic system that proposes a model for application of innovative techniques for data acquisition and analytics in cyber-physical logistic systems
5	Cyber-physical logistics system-based vehicle routing optimization [62, 63]	A smart routing adjustment model that considers the existing road congestion and minimizes the total distribution in cost. Here, static and dynamic models are proposed for traffic information transmission network
6	Quality of logistics services [64]	A research that deals with the customer satisfaction with services and discusses the importance of relation between the logistic companies and customer satisfaction with logistics services

(continued)

Table 7.3 (continued)

S. no.	Application	Purpose
7	CycleLogistics [55]	A research that demonstrates the great potential for the reduction in energy consumption and pollutants caused by urban goods transport, and discusses how to shift from motorized vehicle to bicycles to change the condition of traffic congestion and over pollution
8	Cyber-physical production systems combined with logistic models [65]	A research that discusses the advantages of cyber-physical systems by considering the production planning, controlling and monitoring
9	Hierarchical integrated intelligent logistics system platform [66]	An intelligent system that proposes HIILS platform for the wide spectrum of the city logistics problems
10	Smart city logistics on cloud computing model [56]	An IoT cloud-based smart city logistics model that deals the smart city logistics in terms of sustainable logistics dimensions
11	Parcel industry in the spatial organization of logistics activities [67]	A research that aims to study the location of parcel industry, and its orientation in the spatial organization of logistics activities in the Paris region
12	Smart logistics [57, 68]	A smart logistics system that proposed a smart logistics vehicle management system based on the IoV
13	e-Commerce logistics in supply chain management [69]	A research that presents the state-of-the-art e-Commerce logistics in supply chain management from practice perspective
14	RFID-based data mining for e-Logistics [70]	An intelligent system that combines RFID for data acquisition, data mining for knowledge discovery and enterprise applications in the field of e-Logistics
15	e-Commerce logistics based on the gridding management [58, 71]	An intelligent e-Commerce logistics-based gridding management system that deals with physical distribution and its effect on e-Commerce

E-commerce logistics based on the gridding management, via the comparison analysis between the grid and non-grid distribution system in E-commerce logistics.

The structure and level of E-commerce logistics system (ELS) consist of (1) function entity, and (2) management level. The proposed framework is shown in Fig. 7.8. Table 7.3 summarizes various applications designed and implemented for e-Logistics.

7.3 Summary

This chapter discusses the applications of predictive computing in real life. It deals with the concept of how and where we can explore predictive computing and deploy the applications. The discussion is concentrated on three case studies where predictive computing can be applied to various applications, namely; smart mobility, e-Health and e-Logistics.

References

1. Fico (2017a) Who uses predictive analytics? http://www.fico.com/en/predictive-analytics/understanding-predictive-analytics/who-uses-predictive-analytics. Accessed 11 May 2017
2. Fico (2017b) How can predictive analytics help me? http://www.fico.com/en/predictive-analytics/understanding-predictive-analytics/how-can-predictive-analytics-help-me. Accessed 11 May 2017
3. Roberto LA (2017) 6 applications of predictive analytics in business intelligence. https://www.neuraldesigner.com/blog/6_Applications_of_predictive_analytics_in_business_intelligence. Accessed 11 May 2017
4. Scherf C, Wolter F (2016) Electromobility: overview, examples, approaches. Available via GIZ. http://www.sutp.org/files/contents/documents/resources/B_Technical-Documents/GIZ_SUTP_TD15_E-Mobility.pdf. Accessed 11 May 2017
5. Pattanaik V, Mayank S, Gupta PK, Singh SK (2016) Smart real-time traffic congestion estimation and clustering technique for urban vehicular roads. In: Proceedings of 2016 IEEE region 10 conference TENCON, Singapore, pp 3420–3423
6. Milojevic M, Rakocevic V (2013) Distributed vehicular traffic congestion detection algorithm for urban environments. In: Proceedings of 2013 IEEE vehicular networking conference (VNC), Boston, USA, pp 182–185
7. Abishek C, Kumar M, Kumar P (2009) City traffic congestion control in Indian scenario using wireless sensors network. In: Proceedings of fifth IEEE conference on wireless communication and sensor networks (WCSN), Allahabad, India, 2009, pp 1–6
8. Bilal JM, Jacob D (2007) Intelligent traffic control system. In: Proceedings of 2007 IEEE international conference on signal processing and communications (ICSPC 2007), Dubai, United Arab Emirates, pp 496–499
9. Ye N, Wang ZQ, Malekian R, Lin Q, Wang RC (2015) A method for driving route predictions based on hidden Markov model. Math Probl Eng 1–12
10. Malekian R, Kavishe AF, Maharaj BTJ, Gupta PK, Singh G, Waschefort H (2016) Smart vehicle navigation system using Hidden Markov Model and RFID sensors. Wirel Pers Commun 90(4):1717–1742
11. Balaji GN, Anusha S, Ashwini J (2017) GPS based smart navigation for visually impaired using Bluetooth 3.0. IJIR 3(3):773–776
12. Rama MN, Sudha PN (2016) Smart navigation system for visually challenged people. Int J Ind Electro Electr Eng 45–48
13. Digole RN, Kulkarni SM (2015) Smart navigation system for visually impared person. Int J Adv Res Comput Commun Eng 4(7):53–57
14. Lavanya G, Preethy W, Shameem A, Sushmitha R (2013) Passenger bus alert system for easy navigation of blind. In: Proceedings of international conference on circuits, power and computing technologies (ICCPCT), Nagercoil, India, IEEE, pp 798–802

15. Moosavi V, Hovestadt L (2013) Modeling urban traffic dynamics in coexistence with urban data streams. In: Proceedings of 2nd ACM SIGKDD international workshop on urban computing, Chicago, IL, USA, ACM, p 10
16. Al-Taee MA, Khader OB, Al-Saber NA (2007) Remote monitoring of vehicle diagnostics and location using a smart box with global positioning system and general packet radio service. In: Proceedings of international conference on computer systems and applications (AICCSA'07), Amman, Jordan, IEEE/ACS, pp 385–388
17. Sundar R, Hebbar S, Golla V (2015) Implementing intelligent traffic control system for congestion control, ambulance clearance, and stolen vehicle detection. IEEE Sens J 15 (2):1109–1113
18. Minni R (2013) A cost efficient real-time vehicle tracking system. Int J Comput Appl 81 (11):29–35
19. Liu Z, Zhang A, Li S (2013) Vehicle anti-theft tracking system based on internet of things. In: Proceedings of international conference on vehicular electronics and safety (ICVES), Dongguan, China, IEEE, pp 48–52
20. Lee S, Tewolde G, Kwon J (2014) Design and implementation of vehicle tracking system using GPS/GSM/GPRS technology and smartphone application. In: Proceedings of world forum on internet of things (WF-IoT), Seoul, IEEE, pp 353–358
21. Venkatesh H, Perur SD, Jagadish MC (2015) An approach to make way for intelligent ambulance using IoT. Int J Electr Electron Res 3(1):218–223
22. Rao YR (2017) Automatic smart parking system using internet of things (IOT). Int J Eng Technol Sci Res 4(5)
23. Roy A, Siddiquee J, Datta A, Poddar P, Ganguly G, Bhattacharjee A (2016) Smart traffic & parking management using IoT. In: Proceedings of 7th annual information technology, electronics and mobile communication conference (IEMCON), Vancouver, BC, Canada, pp 1–3
24. Lingling H, Haifeng L, Xu X, Jian LX (2011) An intelligent vehicle monitoring system based on internet of things. In: Proceedings of seventh international conference on computational intelligence and security (CIS), Sanya, Hainan, China, pp 231–233
25. Khanna A, Anand R (2016) IoT based smart parking system. In: Proceedings of international conference on internet of things and applications (IOTA), Pune, India, pp 266–270
26. Shah KA, Jha J (2015) Improvement of traffic monitoring system by density and flow control for Indian road system using IoT. Int J Adv Res Innov Ideas Educ 2(3):2176–2182
27. Choosri N, Park Y, Grudpan S, Chuarjedton P, Ongvisesphaiboon A (2015) IoT-RFID testbed for supporting traffic light control. Int J Inf Electron Eng 5(2):102–106
28. Wang P (2013) A framework for agent-based traffic light control. J Commun Comput 10: 713–716
29. Nidhal A, Ngah UK, Ismail W (2014) Real time traffic congestion detection system. In: Proceedings of 5th international conference on intelligent and advanced systems (ICIAS), Kuala Lumpur, Malaysia, pp 1–5
30. ITU (2008) Implementing e-Health in developing countries. https://www.itu.int/ITU-D/cyb/app/docs/e-Health_prefinal_15092008.PDF. Accessed 18 Apr 2017
31. WHO (2016) From innovation to implementation: eHealth in the WHO European region. http://www.euro.who.int/__data/assets/pdf_file/0012/302331/From-Innovation-to-Implementation-eHealth-Report-EU.pdf. Accessed 18 Apr 2017
32. Modi K, Mohanty RB (2015) M-Health: challenges, benefits, ad keys to successful implementation. Available via Infosys, https://www.infosys.com/industries/insurance/white-papers/Documents/health-challenges-benefits.pdf. Accessed 5 May 2017
33. WHO (2016) global observatory for e-Health series. http://www.who.int/goe/en/. Accessed 3 January 2017

34. Gupta PK, Maharaj BT, Malekian R (2016) A novel and secure IoT based cloud centric architecture to perform predictive analysis of users activities in sustainable health centres. Multimed Tools Appl. doi:10.1007/s11042-016-4050-6
35. Baccar N, Bouallegue R (2014) A new web-based e-Health platform. In: Proceedings of 10th international conference on wireless and mobile computing, networking and communications (WiMob), Larnaca, Cyprus, IEEE, pp 14–19
36. Ahmed S, Abdullah A (2011) e-Healthcare and data management services in a cloud. In: Proceedings of high capacity optical networks and enabling technologies (HONET), Riyadh, Saudi Arabia, pp 248–252
37. Liu W, Park EK (2013) e-Healthcare cloud computing application solutions: cloud-enabling characteristics, challenges and adaptations. In: Proceedings of international conference on computing, networking and communications (ICNC), San Diego, USA, pp 437–443
38. Aruna DS, Godfrey WS, Sasikumar S (2016) Patient health monitoring system (PHMS) using IoT devices. IJCSET 7(3):68–73
39. Piliouras T, Yu PLR, Tian X, Zuo B, Yu S, Paulino J, Mei C, Clerger E, Davis D, Sultana N (2013) Electronic health record systems: a current and future-oriented view. In: Proceedings of long island systems, applications and technology conference (LISAT), Farmingdale, NY, USA, IEEE, pp 1–6
40. Hameed RT, Mohamad OA, Hamid OT, Tapus N (2015) Design of e-Healthcare management system based on cloud and service oriented architecture. In: Proceedings of e-Health and bioengineering conference (EHB), Romania IASI, Romania, pp 1–4
41. Mukherjee S, Dolui K, Datta SK (2014) Patient health management system using e-Health monitoring architecture. In: Proceedings of international advance computing conference (IACC), Gurgaon, India, IEEE pp 400–405
42. Hassanalieragh M, Page A, Soyata T, Sharma G, Aktas M, Mateos G, Kantarci B, Andreescu S (2015) Health monitoring and management using internet-of-things (IoT) sensing with cloud-based processing: opportunities and challenges. In: Proceedings of international conference on services computing (SCC), New York, NY, USA, IEEE, pp 285–292
43. Rajput DS, Gour R (2016) An IoT framework for healthcare monitoring systems. Int J Comput Sci Inf Secur 14(5):451–455
44. Aziz K, Tarapiah S, Ismail SH, Atalla S (2016) Smart real-time healthcare monitoring and tracking system using GSM/GPS technologies. In: Proceedings of 3rd MEC international conference on big data and smart city (ICBDSC), Muscat, Oman, pp 1–7
45. Gupta P, Agrawal D, Chhabra J, Dhir PK (2016) IoT based smart healthcare kit. In: Proceedings of international conference on computational techniques in information and communication technologies (ICCTICT), New Delhi, India, pp 237–242
46. Zhang Y, Liu H, Su X, Jiang P, Wei D (2015) Remote mobile health monitoring system based on smart phone and browser/server structure. J Healthc Eng 6(4):717–738
47. Khoi NM, Saguna S, Mitra K, hlund C (2015) Irehmo: an efficient iot-based remote health monitoring system for smart regions. In: Proceedings of 17th international conference on e-Health networking, application & services (HealthCom), Dalian, China, pp 563–568
48. Gómez J, Oviedo B, Zhuma E (2016) Patient monitoring system based on internet of things. Procedia Comput Sci 83:90–97
49. Kumar R, Rajasekaran MP (2016) An IoT based patient monitoring system using raspberry Pi. In: Proceedings of international conference on computing technologies and intelligent data engineering (ICCTIDE), Kovilpatti, India, pp 1–4
50. Keliris AP, Kolias VD, Nikita KS (2013) Smart cards in healthcare information systems: benefits and limitations. In: Proceedings of 13th international conference on bioinformatics and bioengineering (BIBE), Chania, Greece, IEEE, pp 1–4
51. Latha, NA, Murthy, BR, Sunitha U (2012) Smart card based integrated electronic health record system for clinical practice. Editor Pref 3(10)

52. Yeh KH, Lo NW, Wu TC, Yang TC, Liaw HT (2012) Analysis of an eHealth care system with smart card based authentication. In: Proceedings of seventh Asia joint conference on information security (Asia JCIS), Tokyo, Japan, pp 59–61
53. Mwesigwa C (2013) An e-Health tele-media application for patient management. In: Proceedings of IST-Africa conference and exhibition (IST-Africa), Dublin, Ireland, pp 1–7
54. Khayyat M, Awasthi A (2016) An intelligent multi-agent based model for collaborative logistics systems. Transp Res Procedia 12:325–338
55. Wrighton S, Reiter K (2016) Cycle logistics—moving Europe forward. Transp Res Procedia 12:950–958
56. Nowicka K (2014) Smart city logistics on cloud computing model. Procedia Soc Behav Sci 151:266–281
57. Sharma N, Chauhan N, Chand N (2016) Smart logistics vehicle management system based on internet of vehicles. In: Proceedings of fourth international conference on parallel, distributed and grid computing (PDGC), Shimla, India, pp 495–499
58. Jianyu W, Runtong Z, Jun W, (2015) Research on the distribution system in e-Commerce logistics based on gridding management. In: Proceedings of international conference on logistics, informatics and service sciences (LISS), Barcelona, Spain, pp 1–5
59. Domingues ML, Reis V, Macário R (2015) A comprehensive framework for measuring performance in a third-party logistics provider. Transp Res Procedia 10:662–672
60. He M, Xie J, Wu X, Hu Q, Dai Y (2016) Capability coordination in automobile logistics service supply chain based on reliability. Procedia Eng 137:325–333
61. Frazzon EM, Dutra ML, Vianna WB (2015) Big data applied to cyber-physical logistic systems: conceptual model and perspectives. Braz J Oper Prod Manag 12(2):330–337
62. Lai M, Yang H, Yang S, Zhao J, Xu Y (2014) Cyber-physical logistics system-based vehicle routing optimization. J Ind Manag Optim 10(3):701–715
63. Ayed AB, Halima MB, Alimi AM (2015) Big data analytics for logistics and transportation. In: Proceedings of 4th international conference on advanced logistics and transport (ICALT), Valenciennes, France, pp 311–316
64. Meidutė-Kavaliauskienė I, Aranskis A, Litvinenko M (2014) Consumer satisfaction with the quality of logistics services. Procedia Soc Behav Sci 110:330–340
65. Seitz KF, Nyhuis P (2015) Cyber-physical production systems combined with logistic models—a learning factory concept for an improved production planning and control. Procedia CIRP 32:92–97
66. Adamski A (2011) Hierarchical integrated intelligent logistics system platform. Procedia Soc Behav Sci 20:1004–1016
67. Heitz A, Beziat A (2016) The parcel industry in the spatial organization of logistics activities in the Paris region: inherited spatial patterns and innovations in urban logistics systems. Transp Res Procedia 12:812–824
68. Bärwald W, Baumann S, Keil R, Leitzke C, Richter K (2007) "Smart logistics"-usage of innovative information and communication technologies in production logistics. In: Proceedings of 3rd European workshop on RFID systems and technologies (RFID SysTech), Duisburg, Germany, pp 1–9
69. Yu Y, Wang X, Zhong RY Huang GQ (2016) e-Commerce logistics in supply chain management: practice perspective. Procedia CIRP 52:179–185
70. Wang Y, Yu Q, Wang K (2013) RFID based data mining for e-Logistics. In: Proceedings of international conference on e-Business (ICE-B), Reykjavík, Iceland, pp 1–8
71. Sun T, Xue D, (2015) E-commerce logistics distribution mode research. In: Proceedings of international conference on computational intelligence and communication technology (CICT), Ghaziabad, UP, India, IEEE, pp 699–702

Appendix
Datasets

A large number of datasets are freely available on the Internet that can be used in experiments related to Predictive Computing and Information Security. We have provided a list of commonly used dataset and their links below.

- **Educational Process Mining (EPM): A Learning Analytics Dataset**
  http://archive.ics.uci.edu/ml/datasets/Educational+Process+Mining+%28EPM%29%3A+A+Learning+Analytics+Data+Set

  Data Set Characteristics: Multivariate, Sequential, Time-Series
  Attribute Characteristics: Integer
  Associated Tasks: Classification, Regression, Clustering

This dataset contains the students' time series of activities during six sessions of laboratory sessions of the course of digital electronics. There are six folders containing the students data per session. Each 'Session' folder contains up to 99 CSV files each dedicated to a specific student log during that session. The number of files in each folder changes due to the number of students present in each session. Each file contains 13 features. The experiments have been carried out with a group of 115 students of first-year, undergraduate engineering major of the University of Genoa.

- **Heart Disease Dataset**
  http://archive.ics.uci.edu/ml/datasets/Heart+Disease

  Data Set Characteristics: Multivariate
  Attribute Characteristics: Categorical, Integer, Real
  Associated Tasks: Classification

This database contains 76 attributes, but all published experiments refer to using a subset of 14 of them. The 'goal' field refers to the presence of heart disease in the patient.

P.K. Gupta et al., *Predictive Computing and Information Security*,
DOI 10.1007/978-981-10-5107-4

- **Fertility Dataset**
  http://archive.ics.uci.edu/ml/datasets/Fertility

  Data Set Characteristics: Multivariate
  Attribute Characteristics: Real
  Associated Tasks: Classification, Regression

100 volunteers provide a semen sample analyzed according to the WHO 2010 criteria. Sperm concentration is related to socio-demographic data, environmental factors, health status and life habits.

- **Online News Popularity Dataset**
- http://archive.ics.uci.edu/ml/datasets/Online+News+Popularity

  Dataset Information:

  Data Set Characteristics: Multivariate
  Attribute Characteristics: Integer, Real
  Associated Tasks: Classification, Regression

Online News Popularity dataset summarizes a heterogeneous set of features about articles published by Mashable in a period of 2 years. The goal is to predict the number of shares in social networks (popularity). This dataset does not share the original content but some statistics associated with it. The original content be publicly accessed and retrieved using the provided urls.

- **Appliances energy prediction Dataset**
  http://archive.ics.uci.edu/ml/datasets/Appliances+energy+prediction

  Data Set Characteristics: Multivariate, Time-Series
  Attribute Characteristics: Real
  Associated Tasks: Regression

Experimental data used to create regression models of appliances energy use in a low energy building. The dataset is at 10 min. for about 4.5 months. The house temperature and humidity conditions were monitored with a ZigBee wireless sensor network. Each wireless node transmitted the temperature and humidity conditions around 3.3 min. Then, the wireless data was averaged for 10 minutes periods. The energy data was logged every 10 minutes with m-bus energy meters. Weather from the nearest airport weather station (Chievres Airport, Belgium) was downloaded from a public dataset from Reliable Prognosis (rp5.ru), and merged together with the experimental datasets using the date and time column. Two random variables have been included in the dataset for testing the regression models and to filter out non predictive attributes (parameters).

- **Bank Marketing Dataset**
  http://archive.ics.uci.edu/ml/datasets/Bank+Marketing

  Data Set Characteristics: Multivariate
  Attribute Characteristics: Real
  Associated Tasks: Classification

The data is related with direct marketing campaigns (phone calls) of a Portuguese banking institution. The classification goal is to predict if the client will subscribe a term deposit (variable). The marketing campaigns were based on the phone calls. Often, more than one contact to the same client was required, in order to access if the product (bank term deposit) would be ('yes') or not ('no') subscribed.

- **Numerical Weather Prediction dataset**
  https://www.ncdc.noaa.gov/data-access/model-data/model-datasets/numerical-weather-prediction

Numerical Weather Prediction (NWP) data are the form of weather model data we are most familiar with on a day-to-day basis. NWP focuses on taking current observations of weather and processing these data with computer models to forecast the future state of weather. Knowing the current state of the weather is just as important as the numerical computer models processing the data. Current weather observations serve as input to the numerical computer models through a process known as data assimilation to produce outputs of temperature, precipitation and hundreds of other meteorological elements from the oceans to the top of the atmosphere.

- **Census-Income (KDD) Dataset**
  https://archive.ics.uci.edu/ml/datasets/Census-Income+%28KDD%29

  Data Set Characteristics: Multivariate
  Attribute Characteristics: Categorical, Integer
  Associated Tasks: Classification

Census-Income (KDD) data set contains weighted census data extracted from the 1994 and 1995 current population surveys conducted by the U.S. Census Bureau. The data contains 41 demographic-and employment-related variables. The instance weight indicates the number of people in the population that each record represents due to stratified sampling.

- **URL Reputation Dataset**
  https://archive.ics.uci.edu/ml/datasets/URL+Reputation

  Data Set Characteristics: Multivariate, Time-Series
  Attribute Characteristics: Integer, Real
  Associated Tasks: Classification

Anonymized 120-day subset of the ICML-09 URL data containing 2.4 million examples and 3.2 million features. A label of +1 corresponds to a malicious URL and −1 corresponds to a benign URL.

- **Detect Malacious Executable(AntiVirus) Dataset**
  http://archive.ics.uci.edu/ml/datasets/Detect+Malacious+Executable%28AntiVirus%29

  Data Set Characteristics: Multivariate
  Attribute Characteristics: Real
  Associated Tasks: Classification

Dataset contains training file with 100+ non malicious examples and 250+ malicious samples. Non-Malicious dataset is represented by +1 while Malicious dataset is represented by −1 as label.

- **KDD Cup 1999 Data**
  http://kdd.ics.uci.edu/databases/kddcup99/kddcup99.html

This is the data set used for The Third International Knowledge Discovery and Data Mining Tools Competition, which was held in conjunction with KDD-99 The Fifth International Conference on Knowledge Discovery and Data Mining. The competition task was to build a network intrusion detector, a predictive model capable of distinguishing between 'bad' connections, called intrusions or attacks, and 'good' normal connections. This database contains a standard set of data to be audited, which includes a wide variety of intrusions simulated in a military network environment

- **Phishing Websites Dataset**
  https://archive.ics.uci.edu/ml/datasets/phishing+websites

  Data Set Characteristics: N/A
  Attribute Characteristics: Integer
  Associated Tasks: Classification

This dataset collected mainly from: PhishTank archive, MillerSmiles archive, Google's searching operators. One of the challenges faced by our research was the unavailability of reliable training datasets. In fact, this challenge faces any researcher in the field. However, although plenty of articles about predicting phishing websites have been disseminated these days, no reliable training dataset has been published publically, may be because there is no agreement in literature on the definitive features that characterize phishing webpages, hence it is difficult to shape a dataset that covers all possible features

- **Amazon Web Services Public Datasets**
  https://aws.amazon.com/public-datasets/

  AWS hosts a variety of public datasets that anyone can access for free.

Geospatial and Environmental Datasets

- Landsat on AWS: An ongoing collection of satellite imagery of all land on Earth produced by the Landsat 8 satellite.
- Sentinel-2 on AWS: An ongoing collection of satellite imagery of all land on Earth produced by the Sentinel-2 satellite.
- SpaceNet on AWS: A corpus of commercial satellite imagery and labelled training data to foster innovation in the development of computer vision algorithms.
- MODIS on AWS: Select products from the Moderate Resolution Imaging Spectroradiometer (MODIS) managed by the U.S. Geological Survey and NASA.
- Terrain Tiles: A global dataset providing bare-earth terrain heights, tiled for easy usage and provided on S3.
- NAIP: 1 meter aerial imagery captured during the agricultural growing seasons in the continental U.S.
- NEXRAD on AWS: Real-time and archival data from the Next Generation Weather Radar (NEXRAD) network.
- NASA NEX: A collection of Earth science datasets maintained by NASA, including climate change projections and satellite images of the Earth's surface.
- District of Columbia LiDAR: LiDAR point cloud data for Washington, DC.
- EPA Risk-Screening Environmental Indicators: detailed air model results from EPA's Risk-Screening Environmental Indicators (RSEI) model.
- HIRLAM Weather Model: HIRLAM (High Resolution Limited Area Model) is an operational synoptic and mesoscale weather prediction model managed by the Finnish Meteorological Institute.

Genomics and Life Science Datasets

- 1000 Genomes Project: A detailed map of human genetic variation.
- TCGA on AWS: Raw and processed genomic, transcriptomic and epigenomic data from The Cancer Genome Atlas (TCGA) available to qualified researchers via the Cancer Genomics Cloud.
- ICGC on AWS: Whole genome sequence data available to qualified researchers via The International Cancer Genome Consortium (ICGC).
- 3000 Rice Genome on AWS: Genome sequence of 3,024 rice varieties.
- Genome in a Bottle (GIAB): Several reference genomes to enable translation of whole human genome sequencing to clinical practice.

Datasets for Machine Learning

- Common Crawl: A corpus of web crawl data composed of over 5 billion web pages.
- Amazon Bin Image Dataset: Over 500,000 bin JPEG images and corresponding JSON metadata files describing products in an operating Amazon Fulfillment Center.
- GDELT: Over a quarter-billion records monitoring the world's broadcast, print and web news from nearly every corner of every country, updated daily.

- Multimedia Commons: A collection of nearly 100M images and videos with audio and visual features and annotations.
- Google Books Ngrams: A dataset containing Google Books *n*-gram corpuses.
- SpaceNet on AWS: A corpus of commercial satellite imagery and labelled training data to foster innovation in the development of computer vision algorithms.

Regulatory and Statistical Data

- IRS 990 Filings on AWS: Machine-readable data from certain electronic 990 forms filed with the IRS from 2011 to present
- ACS PUMS on AWS: U.S. Census American Community Survey (ACS) Public Use Microdata Sample (PUMS) is available in a linked data format using the Resource Description Framework (RDF) data model
- USAspending.gov on AWS: USAspending.gov database, which includes data on all spending by the federal government, including contracts, grants, loans, employee salaries and more.

Printed in the United States
By Bookmasters